10.50
2938
Scholar's
Bookshelf

DATE DUE			
MAY 06 1987			
MAY 20 1987			

THE ART OF
LIGHT AND COLOR

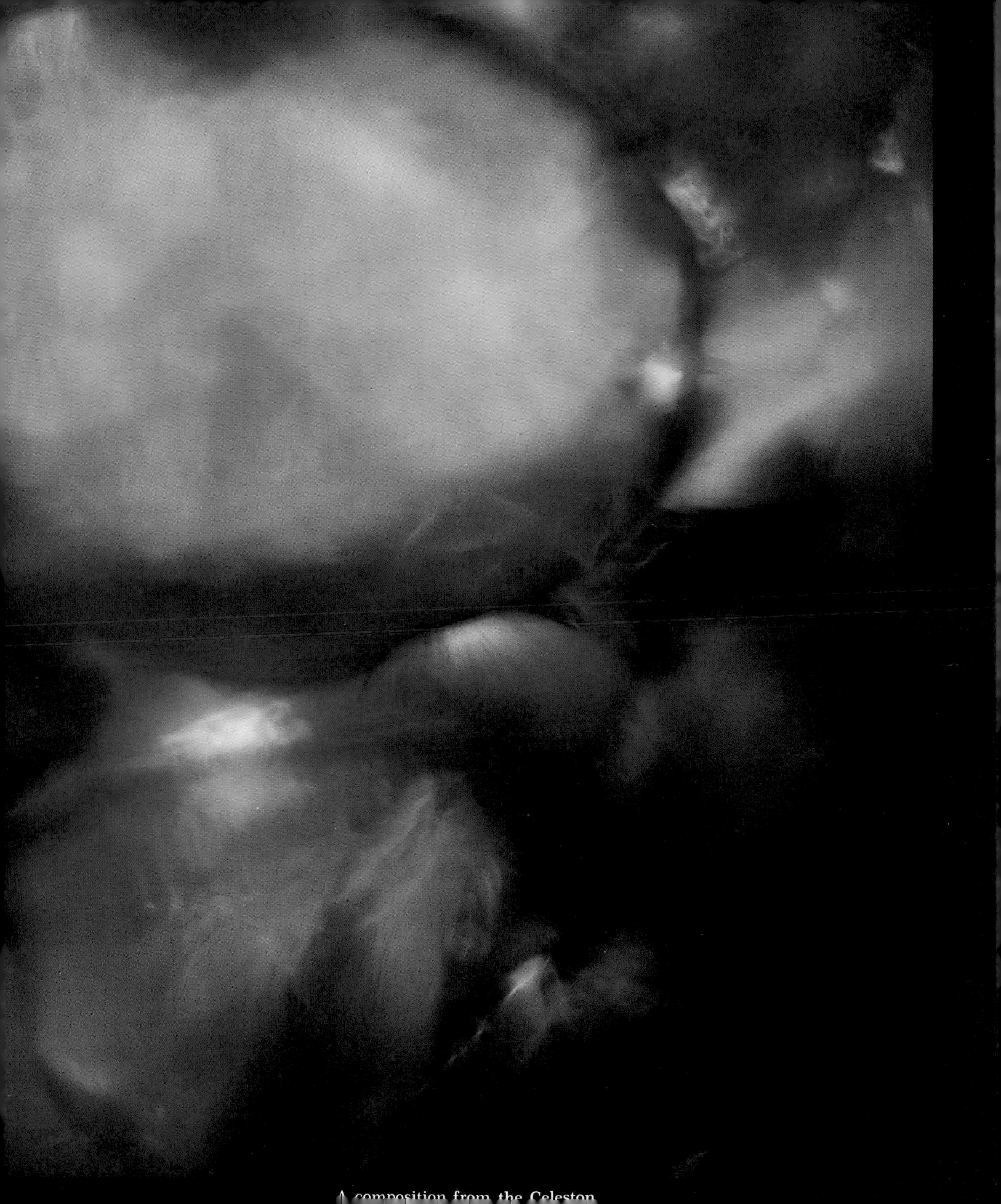

A composition from the Celeston

THE ART OF LIGHT AND COLOR

Featuring Mobile Color Expression, Lumia, Kinetic Light–
with Instructions for the Creation of Dramatic
Color and Light Instruments

by TOM DOUGLAS JONES

With an Introduction by Faber Birren

VNR
VAN NOSTRAND REINHOLD COMPANY
New York / Cincinnati / Toronto / London / Melbourne

VAN NOSTRAND REINHOLD COMPANY REGIONAL OFFICES:
NEW YORK CINCINNATI CHICAGO MILLBRAE DALLAS
VAN NOSTRAND REINHOLD COMPANY INTERNATIONAL OFFICES:
LONDON TORONTO MELBOURNE

LIBRARY OF CONGRESS CATALOG CARD NUMBER: 73-39807

PRINTED BY HALLIDAY LITHOGRAPH CORPORATION
COLOR PRINTED BY PRINCETON POLYCHROME PRESS
BOUND BY THE BOOK PRESS

PUBLISHED BY VAN NOSTRAND REINHOLD COMPANY
450 WEST 33RD STREET, NEW YORK, N. Y. 10001
PUBLISHED SIMULTANEOUSLY IN CANADA BY
VAN NOSTRAND REINHOLD LIMITED
1 3 5 7 9 11 13 15 16 14 12 10 8 6 4 2

ACKNOWLEDGMENTS

To teachers, students, colleagues, and friends—too numerous to name—for their interest and encouragement; to Dr. Roberto A. Guatelli for his assistance in building the present apparatus; to Faber Birren for his guidance in the preparation and writing of this book; to Col. (ret.) B. D. Pile for his collaboration in the production of the photographs; to Margaret Holton for her perceptive editing; and to my wife, Maryruth, for her inspiration and patience through the years.

COLOR PLATES

Contents

Introduction

Tom Douglas Jones was born, raised, and educated in mid-western America. The writer of this introduction heard of his work and met with him around 1948. The two have been in close touch ever since.

Jones has been a pioneer in the art of light and color. His first color projection device, which he called the Symphochrome, was made and demonstrated in 1938. Since then he has continued his creative research, and the results of many years of devotion are presented in this book.

Jones has essentially been an artist, educator, and museum director. He has studied variously at New York University, Columbia University, the studio of Paul Bornet in Paris, and the Beaux-Arts School in Fontainebleau. He has received degrees at the University of Kansas and the University of Iowa. He became Doctor of Fine Arts at Bethany College in Kansas.

In his earlier years, Jones worked as an illustrator and designer in Kansas City, Chicago, and New York. He taught design for several years at the University of Kansas, where he also engaged actively in light and color experiments, using original instruments as part of his teaching procedures. Later he became research professor at Long Island University.

In 1945, however, Jones left his college assignments to join International Business Machines (IBM). Later he worked directly under the senior Thomas J. Watson, heading the Department of Arts and Science, supervising the cataloging and exhibiting of a broad art and science collection, traveling overseas to study museum architecture and administration, and managing the well-known IBM Gallery on 57th Street in New York.

Now residing in Colorado, Jones maintains an elaborate and well-equipped studio in which his light and color instruments are undergoing constant improvement.

To Jones, light and color represent an art form, not a novelty. As he states in Chapter II, there is a rich tradition dating back to the work of Castel of France, Rimington of Great Britain, and Wilfred of the United States. To the list should be added Tom Douglas Jones. This present book, built around a lifetime of effort, is unique and one of the very few ever written on this distinct art.

While the art of light and color has been progressing through the years, sudden impetus followed the psychedelic age, as reviewed in Chapter III. The discotheque, light show, electric circus made a carnival of light, color, and sound phenomena, and brought in strobe lights, abstract motion pictures, mirrors, overhead liquid projectors, polarized and fluorescent color effects, plus a roar of hi-fi sounds. There is excitment in all this, particularly to a young generation. There is novelty, ingenuity, and out of it has come a new host of creative talent, which Jones has fairly acknowledged.

But the more profound art endures and keeps advancing into the future. With the instruments described herein, the Colortron, Sculptachrome, Chromaton, Celeston, a great visual "music" can be composed and played. Teacher and student can be enlightened on the magic of light and color, mediums quite distinct from pigment and dye. Consoles like those of an organ can be used to creative ends, to call forth stirring responses from the eye and emotions.

Finally, Jones has made the book a valuable reference source on the encompassing art of mobile color, kinetic color, lumia. Only a man of his long and experienced background could possibly do as well.

FABER BIRREN

1 An Independent Art

Light and color—mobile color, kinetic color, lumia—as an *independent* art is of comparatively recent development. Throughout history, however, there has always been strong association between color and music. As this book will point out, the author agrees with most modern exponents of color and light that any kinship between color and sound or color and music is purely a matter of human emotional feeling. No literal or scientific correlation between the vibrations of color and the vibrations of sound has ever been convincingly established, despite numerous attempts. It is to be submitted that music has an emotional quality that is akin to the appeal of color. In arts like painting, sculpture, architecture, the dance, form is vital. A certain intellectual interpretation is needed. Beauty lies in a careful blending of many elements, design, proportion, movement, all of which please or displease as the artist wishes or as the viewer interprets. But music and color require practically no effort, merely attention, to enjoy. Both have a primitive appeal and flow readily over the dikes of the brain to innundate the emotions.

This likeness of color to music and music to color has been noted by many musicians and artists. Some composers have, in an abstract way, tried to ally the two arts spiritually. Others have been quite deliberate in their efforts to make color a definite part of music, to assign scales to it, and to "play" it as one might an organ. Still others have looked upon color as an independent art, devoid of sound, but none the less "melodious."

Terms from the two arts—color, music—have been freely borrowed and exchanged. Klein, in his remarkable book *Colour-Music, The Art of Light,* writes, "Musicians have appropriated the word color principally to describe the sensuous charm of their art. Hue is used to denote the shifts in effect that followed various changes in timbre. Tone-color is a synonym for timbre. In truth, this quality in music (timbre), aside from what are called intensity and pitch, is easily associated with hue. For color has timbre, fullness, delicacy, volume, softness. The voice of a coloratura soprano suggests a broad flow of color."

In a phenomenon known as *synesthesia,* or color-hearing—to be treated in further detail in chapter 10—many persons are found who associate certain hues with certain sounds. Why, no one seems to know. But these associations, natural to the psychic make-up of the individual, are for the most part fixed in his character. They are innate and apparently have little to do with memory or experience.

To distinguish between color and light as an art that is independent of sound and music, it will be noted that two colliding viewpoints have encountered each other over the years. First are those who have attempted to ally hue with sound and to develop color scales comparable to music scales. On the other hand are those who have seen color as an independent art, mobile, emotional, and apart from sound. In 1875 H. R. Haweis wrote: "The only possible rival to sound as a vehicle of pure emotion is color. . . . Here I will express my conviction that a color-art exactly analogous to the sound-art of music is possible and is amongst the arts which have to be traversed in the future, as sculpture, architecture, paintings, and music have been in the past."

Thus men like Thomas Wilfred (see next chapter) struggled to be pioneers. M. Luckiesh, a leading authority on lighting, accurately predicted that scientists rather than musicians would one day do much to develop color and light as a unique art.

The idea has warmed the fancies of men for many centuries. Aristotle himself said, "Colors may mutually relate like musical concords for their pleasantest arrangement; like those concords mutually proportionate."

The mighty Sir Isaac Newton favored the alliance. He wrote, "Considering the lastingness of the emotions excited in the bottom of the eye by light, are they not of a vibratory nature? Do not the most refrangible rays excite the shortest vibrations—the least refrangible the largest? May not the harmony and discord of colors arise from the proportions of the vibrations propagated through the fibers of the optic nerve into the brain, as the harmony and discord of sounds arise from the proportions of the vibrations of the air?"

The answer to Newton's question today would be *no*. His color scale was as follows: red for note *C*, orange for *D*, yellow for *E*, green for *F*, blue for *G*, indigo for *A*, and violet for *B*.

1-1. Diagrams for Newton's original *Opticks,* including the first known color circle.

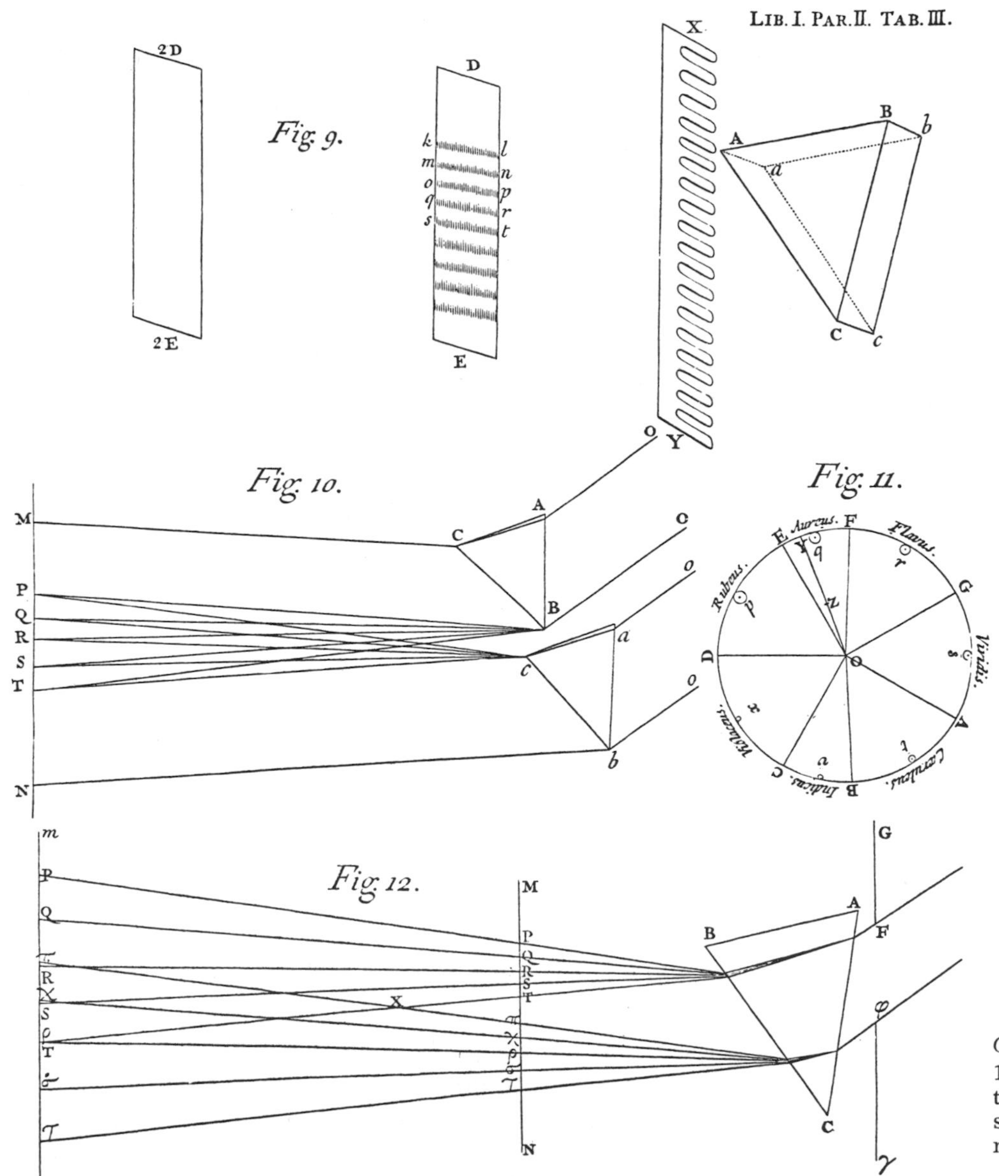

Opposite:
1-2. Newton used these diagrams to illustrate that a prism would separate the colors that existed in natural light.

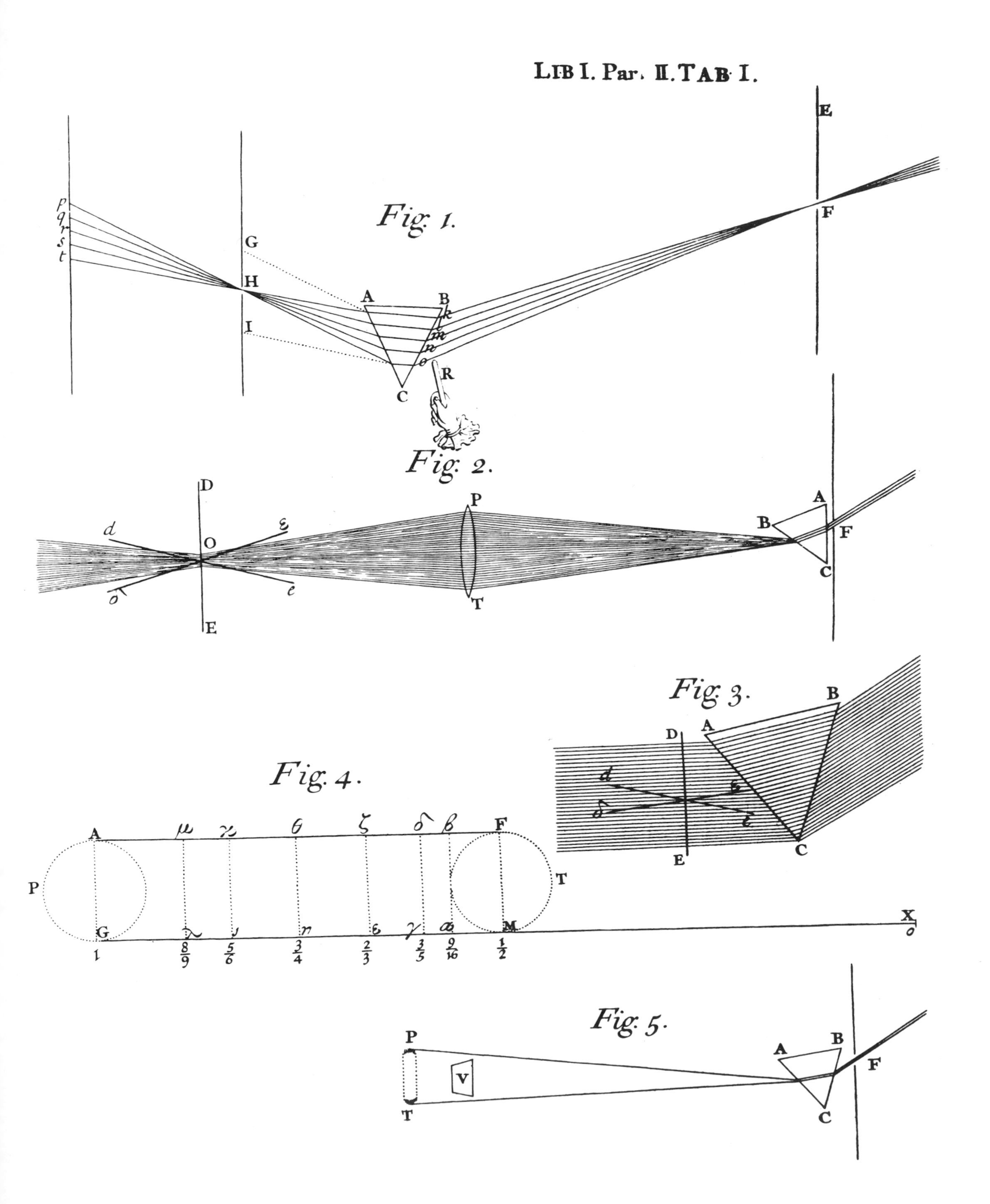
Fig. 1.
p
q
r
s
t
G
H
I
A
B
k
m
n
o
R
C
E
F
Fig. 2.
D
O
E
d
ε
δ
e
P
T
A
B
F
C
Fig. 3.
A
B
D
d
δ
ε
e
E
C
Fig. 4.
A
μ
κ
θ
ζ
δ
β
F
P
T
G
λ
ι
η
ε
γ
α
M
X
o
1
8/9
5/6
3/4
2/3
3/5
9/16
1/2
Fig. 5.
P
T
V
A
B
F
C

1-3. Goethe used a prism like this in his color research (ca. 1792).

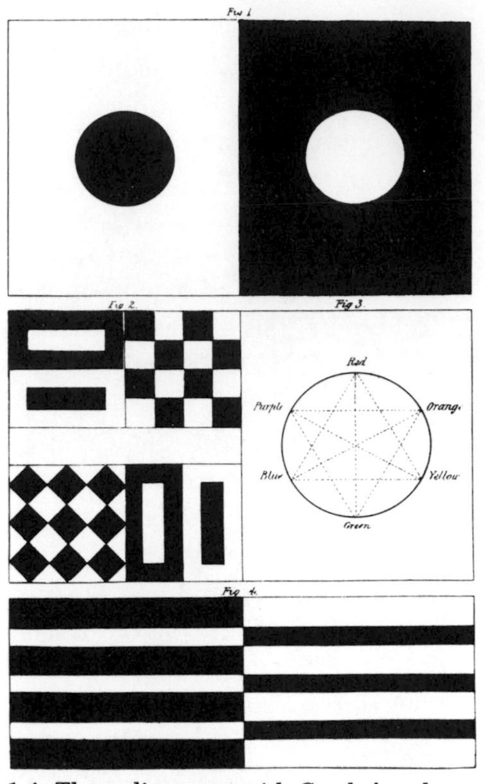

1-4. These diagrams, with Goethe's color circle, appeared in the original edition of his work, *Farbenlehre.*

To continue. Goethe was stoutly opposed. With considerable style he wrote: "Before we proceed to the moral associations of color and the esthetic influences arising from them, we have here to say a few words on its relation to melody. That a certain relation exists between the two has already been felt; this is proved by the frequent comparisons we meet with, sometimes as passing illusions, sometimes as circumstantial parallels. The error which writers have fallen into in trying to establish this analogy we would thus define. Color and sound do not admit of being compared together in any way, but both are referable to a higher formula; both are derivable, although each for itself, from a higher law. They are like two rivers which have their source in one and the same mountain, but subsequently pursue their way under totally different conditions in two totally different regions, so that throughout the whole course of both no two points can be compared. Both are generally elementary effects, acting according to the general law of separation and tendency to union, of undulation and oscillation, yet acting thus in wholly different provinces, in different modes, on different elements and mediums, for different senses."

Thomas Young, the English scientist, was also opposed to an analogy. He wrote, "It appears that any attempt to produce a musical effect from colors must be unsuccessful, or at least that nothing more than a very simple melody could be initiated by them."

M. Chevreul of France, whose great work on simultaneous contrast strongly influenced the schools of Impressionism and Neo-Impressionism, was of like mind with Goethe. "I avow that I cannot perceive those close affinities that several authors, particularly Castel, have said they perceive between sounds and colors. I am ignorant what the future may bring forth, relative to the analogy the senses they respectively affect might present from the point of view of the different kinds of contrast that take place in vision; but at the present time the marked difference between sounds and colors strikes me much more than their generic resemblance."

Helmholtz, the renowned German physicist, after giving sober thought to the matter said abruptly, "I think, for my part, that the comparison might be abandoned."

The opposition thus gained its majority. The art of color must be self-born. Ogden Rood of America and nearly all great colorists after him have failed to see justice in striving to ally color with sound. Albert H. Munsell was silent. Wilhelm Ostwald, the Nobel Prizeman, had earnest objection: "Any attempt to base color harmonies in any way on frequencies is destined to failure at the outset." Color-music, if it is to be written, must fashion its own scales, and these must adhere to the phenomena of vision rather than to spectrophotometry, physics, or fancy laws of vibration.

After all, a careful analysis of the nature of color does not hold much encouragement for a marriage of sound and hue. There are four good reasons for this:

1. The music scale has a natural division into octaves, . . . "inside which," as Ostwald states, "the intermediate notes are repeated in exactly the same relations." This is not true of color. With the gray scale, for example, there are definite terminals in black and white.

2. There is only one "octave" in the spectrum of light. White and black tones of hues do not have wave lengths that differ from their pure originals.

3. One cannot establish an intelligent relationship between color sensation and light frequencies. Red and violet, which are farthest removed in physical character, appear closely related to the eye. Again the waves between 510 and 550 millimicrons (a difference of 40) are all greenish, while the waves between 580 and 590 (a difference of only 10) reveal several hues of yellow-orange that are easy to distinguish. Throughout the spectrum there is no orderly relationship between what the spectrometer records and what the eye sees. In one region a wide span unfolds no great change in hue. In another region large differences exist within a very short span.

4. The eye is unable to distinguish the various frequencies of light seen, for example, in a so-called color-blend. When a chord is struck on a piano, the ear can readily pick out and identify the individual notes. But when red and green lights are thrown upon a common surface the eye sees only yellow, which has no resemblance either to red or green.

2 The Historical Background

Light and color, as an independent art, however, must also embrace references to music, for the early creators and inventors had the two mediums in view. Perhaps the earliest known reference to a color-organ was that of Louis Bertrand Castel (1688–1757), a Jesuit priest eminent in the fields of mathematics, philosophy, and esthetics. In two books, *La Musique en Couleurs* (1720) and *L'Optique des Couleurs* (1740), he described an instrument called the *Clavessin Oculaire*. Castel was a famous and highly respected man of his day. The French philosopher Rousseau was a friend of his and wrote of the "Clavessin Oculaire," as did many following writers and scientists. What Castel did attracted world-wide attention, and in England, during his lifetime, he was made a fellow of the Royal Society.

Castel's first attempt was more or less theoretical. Yet as his idea gained admiration and recognition he apparently devised an instrument of sorts, working with prisms and tapes and employing natural daylight admitted through a window into a darkened room.

Aside from an alliance of colors with notes by the diatonic scale in music, Castel wrote of spectular color effects that today have been unknowingly attempted in the modern discotheques. It is doubtful, of course,

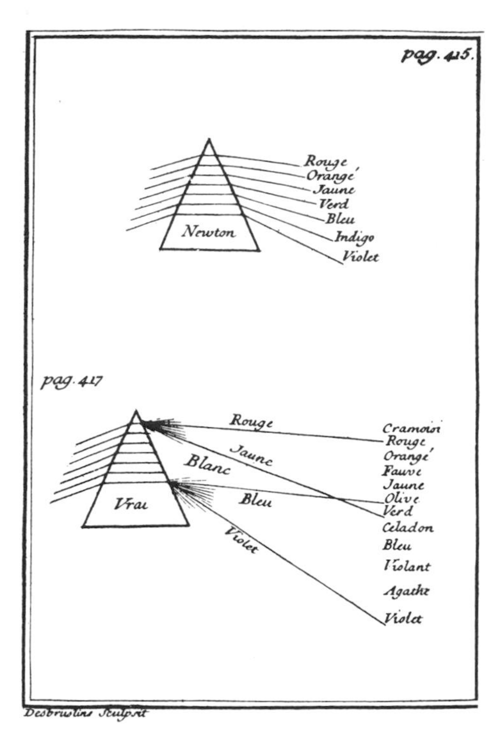

2-1. Louis Castel used this illustration to explain Newton's theory of color (1720).

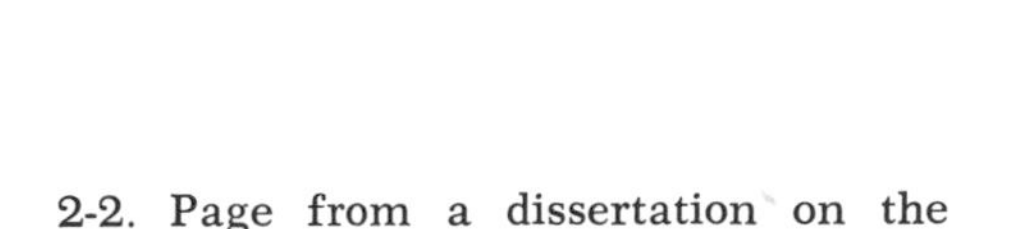

AVDITORES.

Affert nobis lux hodierna tam locupletem gaudendi materiam, aperitque ea tam patentem exſultandi campum, vt, quo certiorem perpetuae felicitatis publicae fiduciam concipere animo nunc licet, et quo magis adverſa omnia a ciuium communi ſalute iam ſunt remota et profligata: eo maiori etiam

A iure

2-2. Page from a dissertation on the work of Castel, Petrograd, 1742.

that he actually achieved what he described, but his imagination well echoes the visions of persons under the influence of hallucinogenic drugs:

"All that visible objects have of magnificence and brilliance can be turned to the profit of the new clavessin. It is susceptible to all manner of embellishments. Gold and azure, metals and enamels, crystals, pearls, diamonds, embroideries, satins, velvets, etc., will not be only ornaments, but will form the body itself of the machine and be as its proper substance. For example, one can form the colors themselves with precious stones or counterfeits of the same color, the reds with garnets and rubies and carbuncles, the greens with emeralds, etc., and what brilliance and splendor a spectacle would possess where one could see appear from all parts and shine like stars, sometimes jacinths, and rubies, and sapphires—all these accompanied with the light of torches in an apartment all hung with mirrors. It would be an infinitely brilliant spectacle as an immobile decoration where everything would be in harmony, but what would it be like if movement and a regular, measured, harmonic, and quick movement animated all, giving it a sort of life? It would be a charm, a glory, a paradise!

"One could perform a play, in which entered human figures, angelic figures, animals, reptiles, etc., or, again, one could demonstrate all the sequence of the elements of Euclid; one can give a play of flowers with variegated flowers, rose for the color of the roses, violet for the violet, etc., so arranged that each touch of the hand would represent a flower-bed and the sequence a mobile diversity of animated flower-beds. All that one can paint one can put into a moving picture, and vice versa, at the will of a clever player of the clavessin. I said that one could make as many color instruments as sound instruments, and one can make them according to a million tastes more different than those of ordinary music. Let all Paris have color clavessins up to 800,000!"

Castel's reference to precious stones has been imitated with bits of glass by later artists. His reference to the elements of Euclid anticipated abstract and non-objective color expression. He was quite a man.

A century was to pass before other men had dreams like those of Castel. In 1844 D. D. Jameson of England described an instrument for the playing of color-music. First of all, the keys of a pianoforte were specially prepared with stains or bits of colored paper. Corresponding swatches were then pasted to the musical score. This not only enabled a person to read music easily, but to see it in terms of color and to have the colors appear themselves.

His instrument had keys connected with a series of twelve round apertures. These apertures were in a chamber holding glass globes filled with translucent liquids. Artificial light sources provided controlled illumination. When the keys of the instrument were struck hues were evolved: "The factors of music and colorific exhibition being thus correlatively fixed, the performance of the one is attended with the other; which has an enchanting effect."

Jameson may have used more logical methods than Castel but the limitations as to artificial illumination were so restricted that one can doubt if his color effects were as enchanting as he claimed.

In America Bainbridge Bishop (1877) built an instrument atop a home organ and with it blended colors on a small screen. These hues were controlled from a keyboard. Both daylight and artificial light were employed. P. T. Barnum had one installed in his home at Bridgeport, Connecticut.

Van Deering Perrine, an American painter, developed two similar gadgets. One consisted of hydraulic pistons that blended liquids in a tank. The others had colors painted on tissues in various blends, these concords being projected in an aperture.

In Australia, Alexander Hector gave concerts with an instrument made of incandescent lamps, Geissler and X-ray tubes. Like Wallace Rimington (see below), his color projections were accompanied by music.

About 1900 E. G. Lind of Baltimore wrote some glowing notes on color-music. He translated national songs into colored diagrams, working out a mathematical scale in which he compared the vibrations of sounds with the vibrations of colors, each figure divisible by the magical number seven.

Plate I. (*A*) THE FIREBIRD, after Thomas Wilfred, 1934. (*B*) Chromation sequence by a student of T. D. Jones, 1940.

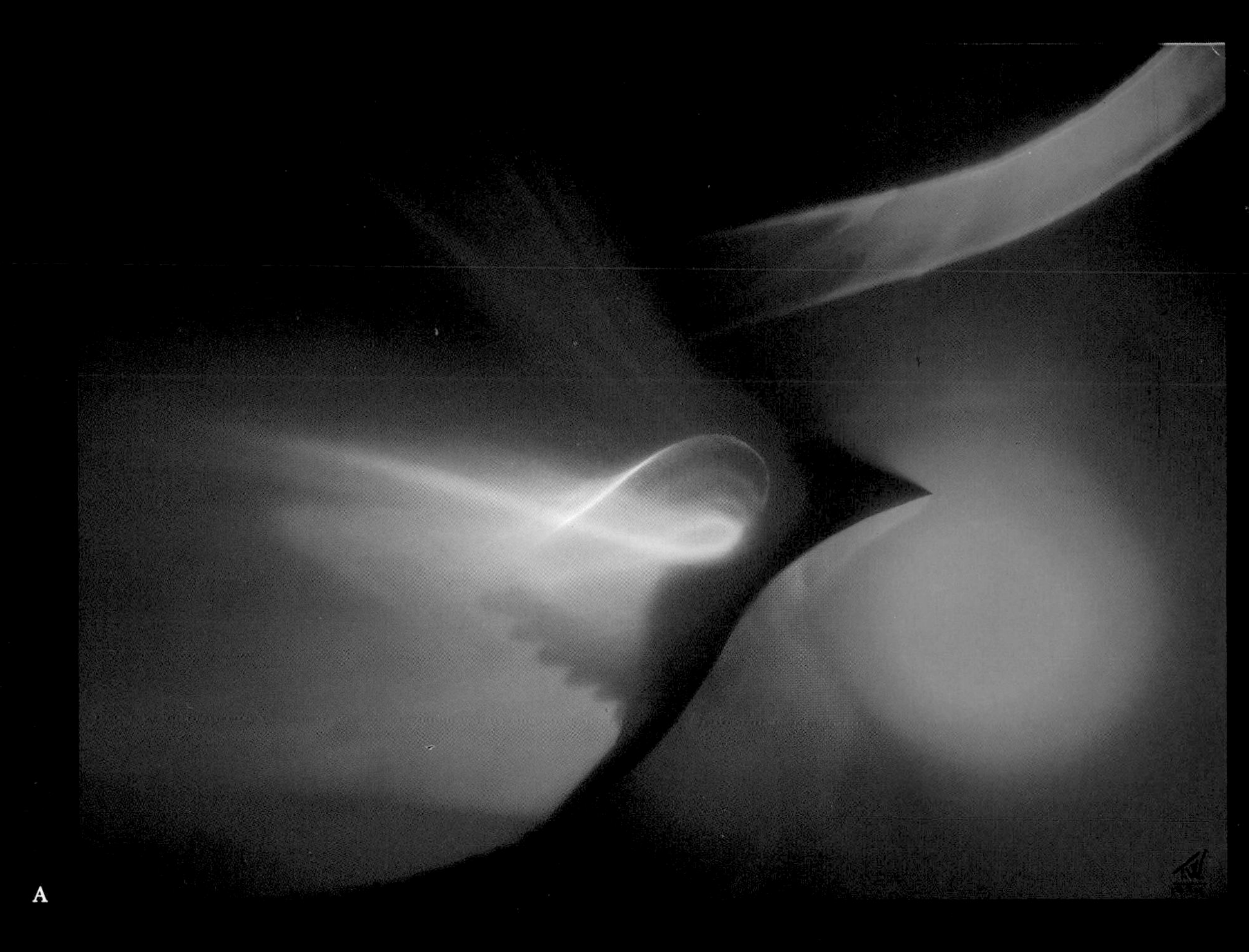

A

B

PLATE I

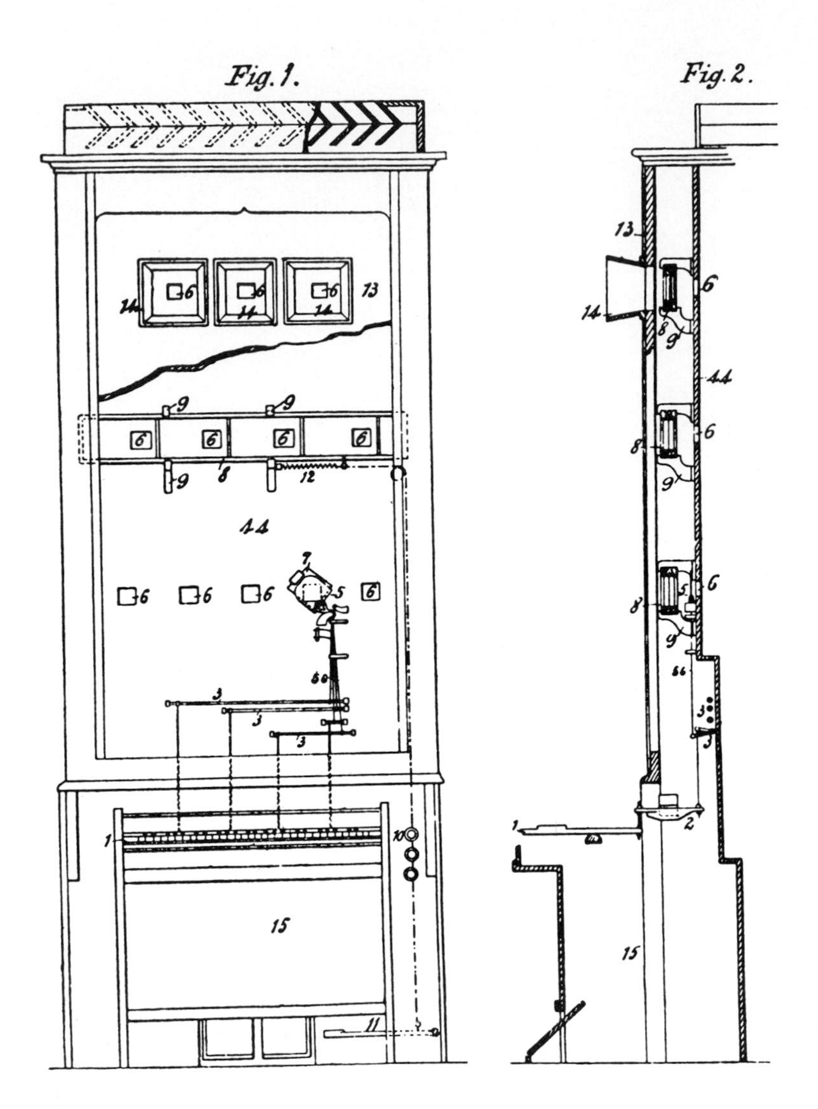

2-3. Diagram, for patent application, of the Color-Organ of A. Wallace Rimington (1893).

Great figures soon appeared. In 1893, A. Wallace Rimington developed a color-organ designed for a joint symphony of music and color. He wrote an important book on his achievements, *Colour-Music, the Art of Mobile Colour*.

His "music" became the rage of London. The instrument he used was elaborately contrived of lights and filters, made possible by developments in electric power and new light sources. His keyboard was similar to the console of an organ. Rimington was trained as an artist. He was inspired by the near-abstract paintings of Turner, in which vivid effects of illumination were achieved, and he saw in colored light a potent new art form. His equipment was expensive and unwieldly; it brought together great arc lamps, lenses, reflectors, masses of cable. Concerts and demonstrations were presented to distinguished audiences. It may have been, however, that his programming of color in direct relation to music (Wagner, Chopin Dvorak) failed to "come off" theatrically. Yet his contribution was an im portant one and inspired many followers.

In America, Mary Hallock-Greenewalt, a pianist, constructed and patented a color-organ and won a gold medal for it at the Sesqui-Centennial International Exposition at Philadelphia in 1926. In her instrument there was a series of lights, controlled by rheostats, which threw hues upon a background. "It is made to stand in the orchestra pit, with the player under the control of the conductor. The instrument is capable of giving forth a light scale conforming with a musical scale." Her theory as to the relationship of color and sound is best expressed in her own words: "Light and color are two entirely separate, distinct, and different things. Light and color can speak as an art alone, as an art of abstract expression made in time succession, but, since the nature of light is that of an accompaniment to all happenings, it will, for some time, in great part be used as such." This remark still holds true. Color and light can be an independent art but such art seems best to lend itself to harmonious color and sound effects at one and the same time, without an attempt at scientific agreement.

One of the most highly regarded men in the field of mobile color has been Thomas Wilfred (1889–1968). His famous "Clavilux" has become synonymous with the general conception of mobile color and light expression. Comprised of spot and flood lights, rheostats, screens and filters, prisms, all under the control of an elaborate console, it is capable of remarkable effects. His term "lumia" has become generic for this form of art.

Wilfred, born in Denmark, came to America and began his experiments in 1905. While his only interest was in color without form, he soon began to supplement this with the introduction of free-flowing forms. His art for the most part was independent of sound, although music was frequently used in conjunction with it. Describing his harmonies he wrote: "The even color which floods the screen is called the accompaniment. In the center of this is the 'solo' figure. This figure may be square, it may be circular, it may be a combination of various figures, as, say, a combination of pyramids, which will turn, and twist, and stretch upwards like arms. The solo figure is always opening and closing, approaching and withdrawing. I have no pet color. The whole spectrum is my favorite. No special color as an especial meaning. Green is generally considered a restful color, but green has a thousand qualities. It may be stirring rather than restful. Blues may mean one thing when applied to a square and another thing when applied to a circle. The key of C major has no especial meaning, but can be made to mean anything that one wishes to make it mean." See Color Plate I for a view of THE FIREBIRD (1934).

The debut of Wilfred's Clavilux was at the Neighborhood Playhouse in New York in 1922. What he had to show held tremendous interest, and he promptly set forth as an impresario, composer, conductor, and performer of the art of color. During 1923–24 he made tours and gave concerts in the United States and Canada. In 1925 he appeared in Paris, London, and Copenhagen. For a while he maintained a small theater and gave public demonstrations in the old Grand Central Palace in New York. After this he maintained an Art Institute of Light in West Nyack, New York. He was honored on many occasions. The Museum of Modern Art in New York gave a permanent home to his Clavilux and a small theater in which the art of lumia was constantly offered. Kinetic lighting is in virtually every recent exhibition devoted to mobile color. Wilfred has been justly remembered as a patriarch and pioneer.

In 1971 the Corcoran Gallery of Art in Washington, D.C., honored Wilfred with a memorial show. In conjunction with it, Donna M. Stein wrote an excellent and beautifully illustrated tribute: *Thomas Wilfred: Lumia, A Retrospective Exhibition*. Assembled were colored instruments, details of his apparatus, scores, drawings, and sketches, and renderings of "An Art Institute of Light," which Wilfred dreamed of but which never became a reality. The Corcoran show was repeated at the Museum of Modern Art in New York City.

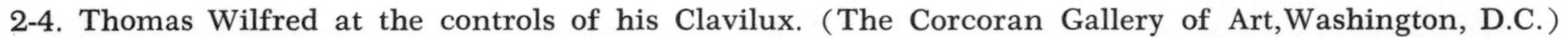

2-4. Thomas Wilfred at the controls of his Clavilux. (The Corcoran Gallery of Art, Washington, D.C.)

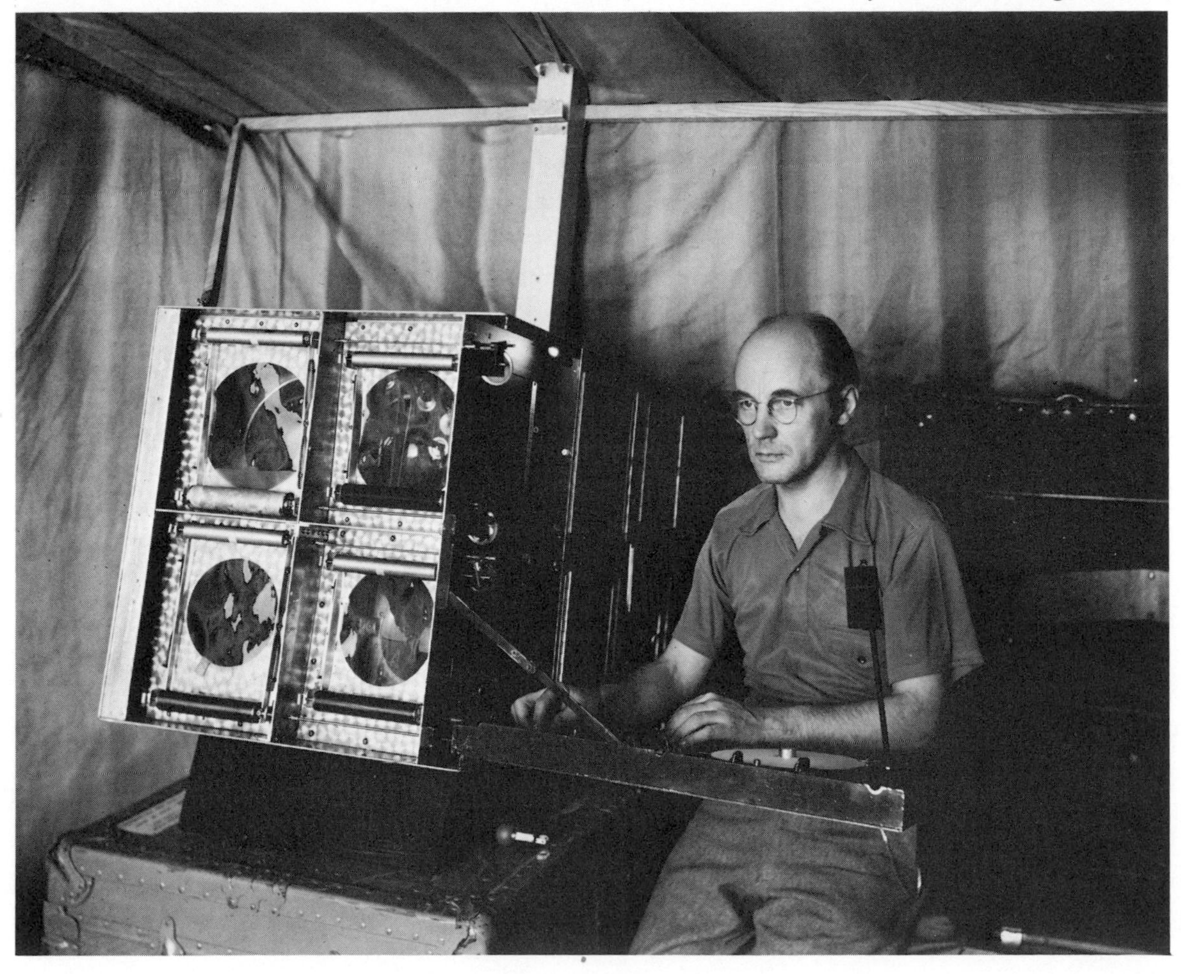

2-5. Still view from Spiral Etude by Thomas Wilfred.

Wilfred's work was seen and admired by such men as Claude Bragdon, Leopold Stokowski, Arthur B. Davies, D. H. Lawrence. He gave command performances for royalty, composed visual settings for "Scheherazade" by Rimsky-Korsakov, did backgrounds for Ibsen's play *The Vikings,* and at one time (1939) created a mobile Rorschach test with a psychologist at the Payne-Whitney Psychiatric Clinic, in New York City, to study schizophrenia.

In the early work of this writer (Jones), Wilfred was duly admired. However, I worked alone, with techniques of my own, and achieved unique results. See Color Plate I for a sequence out of the Chromaton, which dates to around 1940. This mobile composition, done by one of my students, involved both structural form and color.

There have been other admirers of Wilfred. Among them is Earl Reiback, a nuclear engineer who, incidentally, owns many of the instruments, drawings, and other memorabilia from the estate of Thomas Wilfred.

Reiback defines his work as Lumia Aurora. Being technically trained, he has evolved a complex but effectively controlled art form.

The lumia image is created by a programmed optical system and is shaped by a combination of abstract and sculptured forms. These act in a synchronous way to shape images and guide their flow.

The transmission filters, comprising an abstract painting that is individual to each work, are executed on polished glass with high-saturation deep chroma dyes. These dyes, which govern the basic color relationships in the work, are in a highly stable epoxy base, and multiple reflections from the sculptural forms blend the colors and interrelate them for great variety. The image is not random, but rather is programmed for a complete cycle that can last several hours. There may or may not be a music accompaniment.

Reiback's work is shown on Color Plate II. What is seen? To quote *Art News*, "Beautiful colors explode, melt, drift, and dazzle in patterns and movements initiated by the music that is fed into the circuitry of the machine—not merely illustrating the music, but illuminating it, creating an involved orchestration."

In England, John Healey has created a series of mobile geometric abstractions in full color through the use of lenses and mirrors. The forms are less nebulous than regular, like the petals of a flower. They have been applied therapeutically in the University College Hospital of London among disturbed or confined patients. The benefit of such exposure to color is, of course, in the emotional and psychological realm. Here as with the Auroratone motion pictures of Cecil Stokes (see Chapter 9 and Color Plate II) the patient is pleasantly distracted and mentally soothed.

2-6. Tom Douglas Jones at the controls of his Chromaton, 1940.

2-7. Studio and laboratory of Tom Douglas Jones, Long Island University.

In France, Nicolas Schöfer has created what he calls "the Musiscope." Controlled by a keyboard similar to that of an organ, abstract forms in motion are projected on a screen and accompanied by music. A certain measure of manual control is possible. Schöffer conceived of walls of light in architecture in which a continuous play of mobile colored light, harmonized with sound, would set the mood for theatrical productions, opera, ballet—or whatever other human occupation would be visually and emotionally enhanced.

Quite recently Edward Carlton Snyder of New York exhibited a "light brush" that, under fair control, "plays" color to music. Sequences are available in motion pictures and tapes. A New York Light Ensemble has been organized to bring together musicians and color-light projectionists.

Further remarkable work has been done by W. Christian Sidenius of Sandy Hook, Connecticut. (See Color Plate II.) An admirer of Thomas Wilfred, he had engineering training but took a keen interest in mobile color—lumia—around 1945. He has since pursued the art with rare competence, ingenuity, and imagination. He has built a Theater of Light, which seats a few dozen persons, and he conducts regular concerts.

Sidenius has highly sophisticated equipment involving gadgets, electronic devices, and projectors by the score. His compositions may be accompanied by choreography and music. Specially composed music by Tod Dockstader, an electronics expert, and Paul Winter leads to highly original creative expression. Louis Untermeyer, the poet, comments, "Sidenius has done what many of the modern artists would like to do—putting abstractions on canvas instead of static compositions. Many artists try to suggest motion. Sidenius doesn't have to. It's part of his art."

His compositions have been thus described: "Images materialize, drift, streak, snap, explode, blend, separate, pulse with color, and drain to dim grays. They can be wispy, tenuous, hard-edged, and persuasive, always coming and going in dense profusion and simplicity."

In his own words: "Our projectors are all optical consoles and can be performed on like musical instruments. They can give changing color

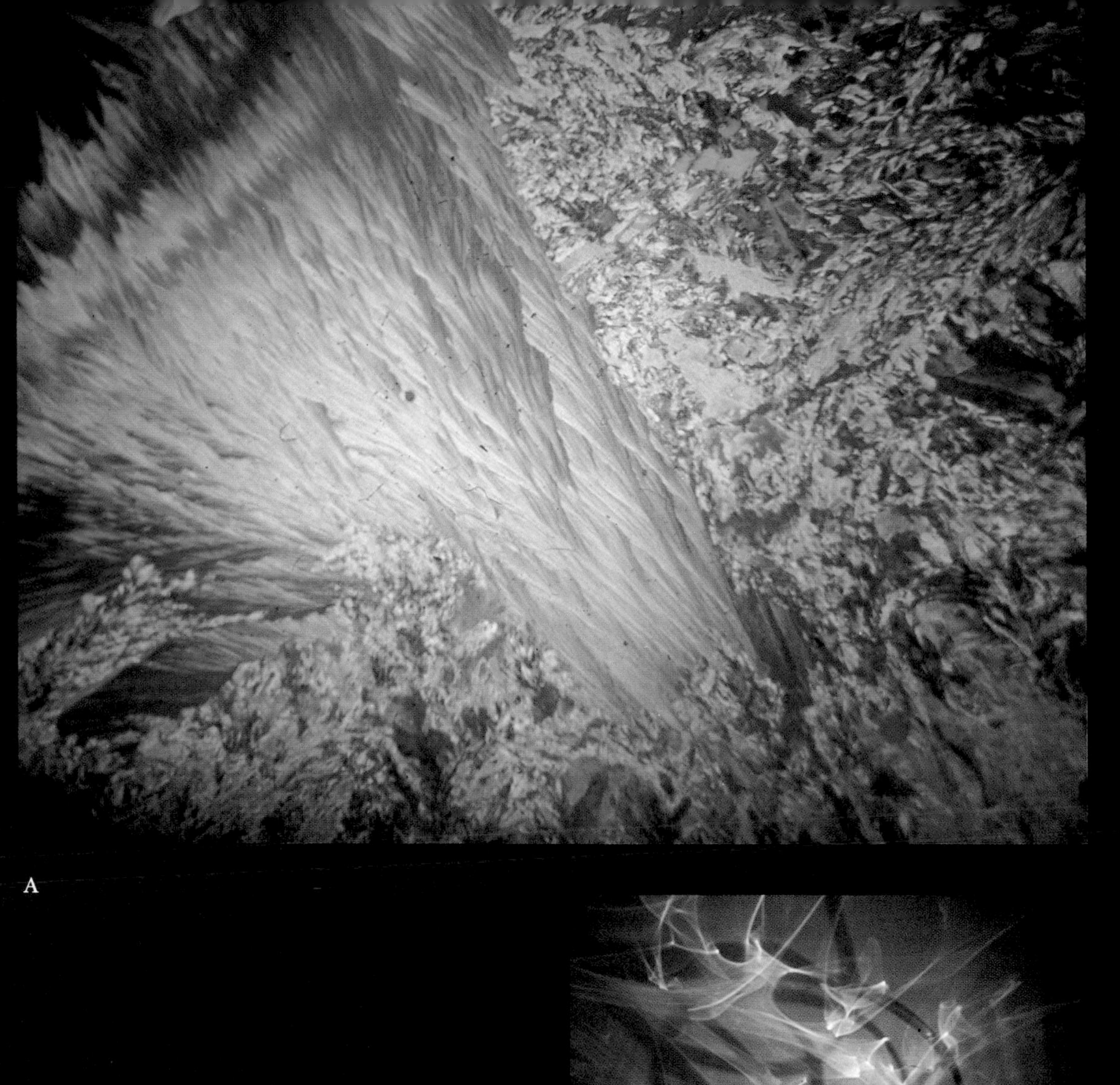

A

B

C

PLATE II

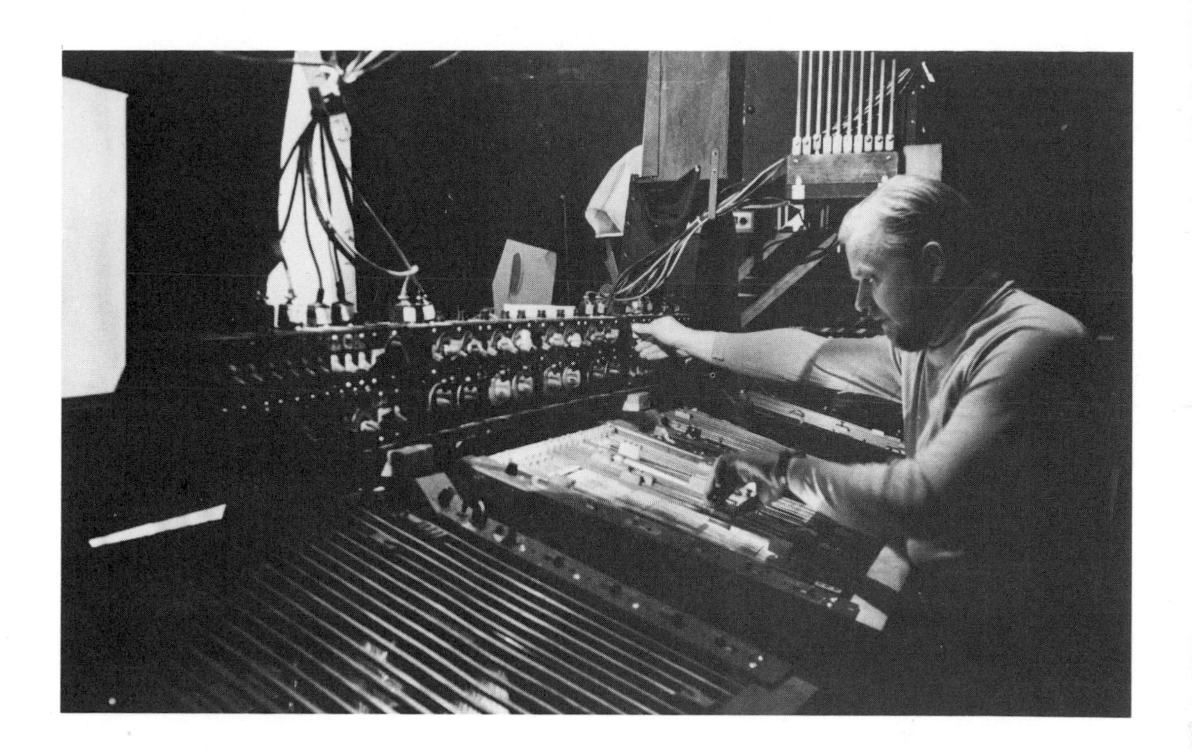

2-8. W. Christian Sidenius at the controls of his color organ.

images that painters cannot and can be choreographed to tell a story. Each figure gives a different type of abstract imagery. Either you look for the visual effect or you fantasize—take your choice."

Sidenius works as an artist in individual lumia sequences. These may be quite varied in effect and quite different in his technical method of presentation. Elaborate preparations may be made. Unlike much mobile color, which tends to be abstract and more or less free-flowing, Sidenius has complete control. Seated at an elaborate console over six feet across, and manipulating projectors, lamps, screens, masks, he follows a taped score and synchronizes his visual patterns with music, much of which is original, composed and arranged for him. His accomplishments are unique, striking, and carry a subtle note of humor.

The art of mobile color has indeed come of age. It has become an independent art. It has met the following prediction that the American physicist and Nobel Prizeman Albert A. Michelson made in 1903: "Indeed, so strongly do these color phenomena appeal to me that I venture to predict that in the not very distant future there may be a color art analogous to the art of sound—a color music, in which the performer, seated before a literally chromatic scale, can play the colors of the spectrum in any succession or combination, flashing on a screen all possible gradations of color, simultaneously or in any desired succession, producing at will the most delicate and subtle modulations of light and color, or the most gorgeous and startling contrasts and color chords! It seems to me that we have here at least as great a possibility of rendering all the fancies, moods, and emotions of the human mind as in the older art."

Plate II. (A) Type of color effect created by Cecil Stokes (see Chapter 9). *(B)* Lumia sequence by Earl Reibeck. *(C)* The mobile color expression of W. Christian Sidenius.

3 This Psychedelic Age

Lumia, the art of light and color, had a sudden resurgence, expansion, and "explosion" during the nineteen sixties, and for two significant reasons. First came the revolt of youth, a sharp break and a full swing away from the amenities and mores of the past. With it was a search for personal freedom. Among numerous other consequences, the sophisticated night club, patronized by well-dressed adults, gave way to the discothèque, the electric circus, frequented by youngsters in dungarees and with bare feet. Rock and roll music, amplified to a cacophonous din, demanded all that the senses could bear—which meant vivid color, flashing light, dizzying motion, stroboscopic vibration. A host of young and ecstatic lumia exponents rushed to the scene, drawn from the fields of art, science, and many other walks of life.

Second came the widespread use of halluciogenic drugs, LSD, mescaline, peyote, the taking of which produced an immediate and startling expansion of the sense of color. Any number of attempts have been made to describe, in words, this heightened and sensuous response to color. Such effort is futile. As Heinrich Klüver wrote in his *Mescal and Mechanisms of Hallucinations*, "It is impossible to find words to describe mescal colors." What is known is that the enlarged visual experience of the drug-taker is closer than anything else to what is seen in kinetic light demonstrations, in pepped up and speeded up abstract lumia.

3-1. The discothèque was a phenomenon of the sixties and was marked by youthful appeal, bright color, and loud sound. (United Press International Photo.)

3-2 One of the most famous discothèques was the Electric Circus in New York where new techniques in electronic sound and color projection were invented. (United Press International Photo.)

The effects of mescaline, and its synthetic counterpart LSD, have been studied over many years and today an extensive literature is available for reference. Klüver mentions that mescal ceremonies and peyote cults have existed over many years. The taking of such mind-expanding drugs is a part of the religion of Indians in southwestern United States; today such cults are also found among Caucasians and other races everywhere and have reached epidemic proportions in many places. First manifestations reveal a heightened sensitivity to color, to fountains and floods of it that outshine the brilliance of gems, northern lights, and all the majestic phenomena of nature. Visions include fanciful appearances of wavy lines, mosaics, ornaments, Oriental carpets, windmills, landscapes, humans and animals, architectural patterns, gratings, fretwork, tunnels, funnels, cones, spirals, on and on.

Aldous Huxley, in *The Doors of Perception,* wrote, "The typical mescaline . . . experience begins with perceptions of colored moving, living geometrical forms. In time, pure geometry becomes concrete, and the visionary perceives, not patterns, but patterned things, such as carpets, carvings, mosaics. These give place to vast and complicated buildings, in the midst of landscapes, which change continuously, passing from richness to more intensely colored richness, from grandeur to deepening grandeur. Heroic figures, of the kind that Blake called 'The Seraphim,' may make their appearance, alone or in multitudes. Fabulous animals move across the scene. Everything is novel and amazing. Almost never does the visionary see anything that reminds him of his own past. He is not remembering scenes, persons, or objects, and he is not inventing them; he is looking on at a new creation."

Anyone who has ever attended a discothèque, an electric circus, a light festival, will know that someone connected with it is familiar with the effects of halluciogenic drugs. Creative and inventive development and expansion of the whole art of light and color may not precisely originate with mescaline and LSD, but surely the results of taking the drugs have provided visions as to what can be accomplished.

3-3. WHIRLWIND OF LOVERS, by William Blake. Fantasies with brilliantly luminous colors. (The Metropolitan Museum of Art. Rogers Fund, 1917.)

What the discothèque and electric circus do is to reverse the procedure of taking drugs. Using flashing lights, bold colors, fluid designs and patterns, thumping sounds like heart beats, roaring music, the real world is blocked out by one of nightmarish fancy—and no drugs need be taken. The senses are overloaded. Inhibitions break down or melt away. Indeed, clinical studies have shown that flashing red lights have been found to induce seizures in epilepsy, while pulsating, stroboscopic lights are hypnotic and can produce headaches, nausea, and minor forms of a nervous breakdown. Much of which is quite appealing to the young!

The art of light and color is in the midst of a new age. Art-Light Shows are being staged everywhere. Public and private museums across America —and in Europe—are inviting the interest and attention of new artists dealing with new art forms. Here follow a few key influences that have taken the work of Thomas Wilfred and others to new domains.

Timothy Leary, whose League of Spiritual Discovery and experiences with LSD were once notorious and led to his arrest and self-imposed exile, in 1966 presented a series of "religious" services in a theater on the lower east side of New York. Color and light projections were visually dominant and were the creative invention of two young people under thirty (at the time), Jackie Cassen and Rudi Stern. It was Leary's view that color and light could cause a sensory disorientation and keener awareness that corresponded to an LSD trip. The two belonged together. The light festival thus was to be part of a modern world of the mind and spirit and could be used as an appropriate stimulus to get people to accept a new order of consciousness.

Quite in another sphere was Fanflashtick, the brain child of Gerd Stern, media poet, and Michael Callahan, engineer, as members of USCO (the US Company). Into a small clear-plastic discothèque, within which were flashing strobe lights, whirling balloons, confetti, taped sounds, entered a number of audience participants. Result: "Lights have a way of making us time-conscious. The flashing allows us to make pattern discriminations within time. The flash of the strobes freezes the movements of people, and people really do move differently from one another. You should see the diffrence in the way the Fanflashtick looks when it's filled with basketball players and how it looks when it's filled with girls from a small Catholic college." (Quoted from an article in *The New York Times Magazine,* May 12, 1968.)

3-4. View of Fanflashtick of Gerd Stern and Michael Callahan.

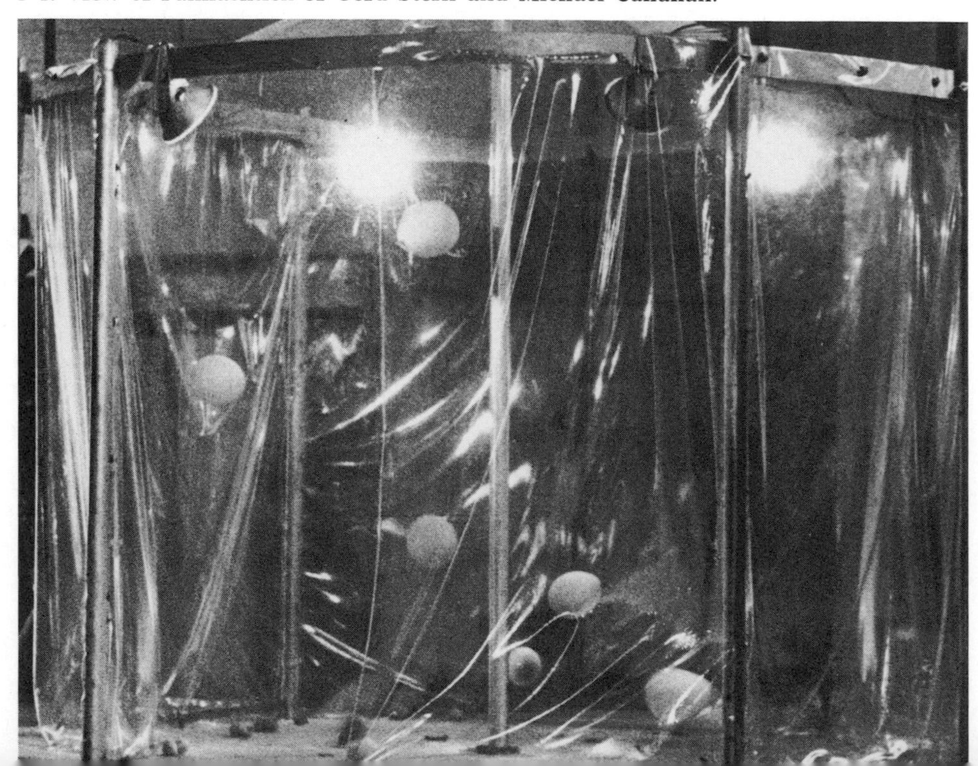

Following a non-too-successful celebration called "Nine Evenings of Theater and Engineering" in New York, 1966, new artists dedicated to light and color emerged. Here again, as with Leary, lumia became the rage of the young at heart, artists or otherwise. This led to the formation of an organization known as EAT, "Experiments in Art and Technology." Led by J. Wilhelm Klüver, EAT has had high-level prestige and has been supported by museums and endowed by such sober companies at AT&T, Xerox, and IBM. Light shows and kinetic shows have been held. Far from the Greenwich Village aura—and the discothèque—EAT hopes to create harmony between science and art. It has held meetings in which scientists have talked to artists—and the converse.

On April 9, 1968, the Inter-Society Color Council held a meeting in New York at which color and light demonstrations were made by top experts. Presenting the program was Klüver of EAT. Among others was Earl Reiback, a nuclear engineer mentioned in Chapter 2. With various projectors he showed advanced techniques for lumia in which abstract patterns, painted with transparent chemicals, were bombarded with laser beams and gamma rays to form weird effects and contortions. Reiback has well introduced science into mobile art.

Featured also was Tony Martin of New York University. He showed the possibility of notating and programming light and sound, having to his credit environments created for the Electric Circus and a full concert at Carnegie Hall.

Today, at the low end, are scores of gadgets, lights, filters, hi-fi components that turn light into color, all available from mail-order houses and electric supply shops. Light shows become commonplace in barrooms, parties, bazaars, church and synagogue socials—not to overlook abstruse Zen rituals in underground dens. Light and color turn people on to ends unknown—and perhaps often meaningless.

At the high end are the apostles and followers of Castel, Rimington, and Wilfred. Light and color suggest time and space, movement, infinity. As a later chapter, on "Color and Psychotherapy," will submit, the mediums of colored light, abstract form, and music strike profound chords in human beings. And in days when men live more and more in environments of their own creation, when they are subjected not only to the fears and bewilderment fostered by a complex civilization, but to the hazards of what is known as "sensory deprivation," it seems quite certain that the art of mobile color will very much come to the rescue.

3-5. GROWTH CYCLES, a painting by Isaac Abrams meant to interpret psychedelic experience.

4 Additive Color Mixture

Around 1916 Wilhelm Ostwald, the great German color theorist and Nobel Prizeman in chemistry, made the remarkable observation that the colors seen by the human eye were of distinctly different types. Some were related and others were unrelated. Physicists of the stature of Hermann von Helmholtz had missed the point, and so had other scientists. Related colors were those associated with the reflection or transmission of material substances. Another term for them, later devised, was *surface* colors. Virtually always they contained some measure of black. They belonged to most of the things seen in this world.

Unrelated colors, however, were those associated with light itself. Here an alternate term became *film* colors. (David Katz, psychologist, had a third term, *volume* colors, which occupied three-dimensional space like smoke or glass jars of dyed liquid.) Between related colors and unrelated colors, there was one big difference: the related colors were always localized, and they were seldom pure. Wrote Ostwald, "It is impossible to obtain *Brown, Olive Green, Grey-Blue,* or in fact any dull color whatsoever, by mixing a *Spectral Color* with *White.*"

The art of light and color uses unrelated colors (film colors) essentially, colors such as seen in the sky, in the rainbow, during dawn and dusk. They are always pure. Consider the following: if black paint is added to

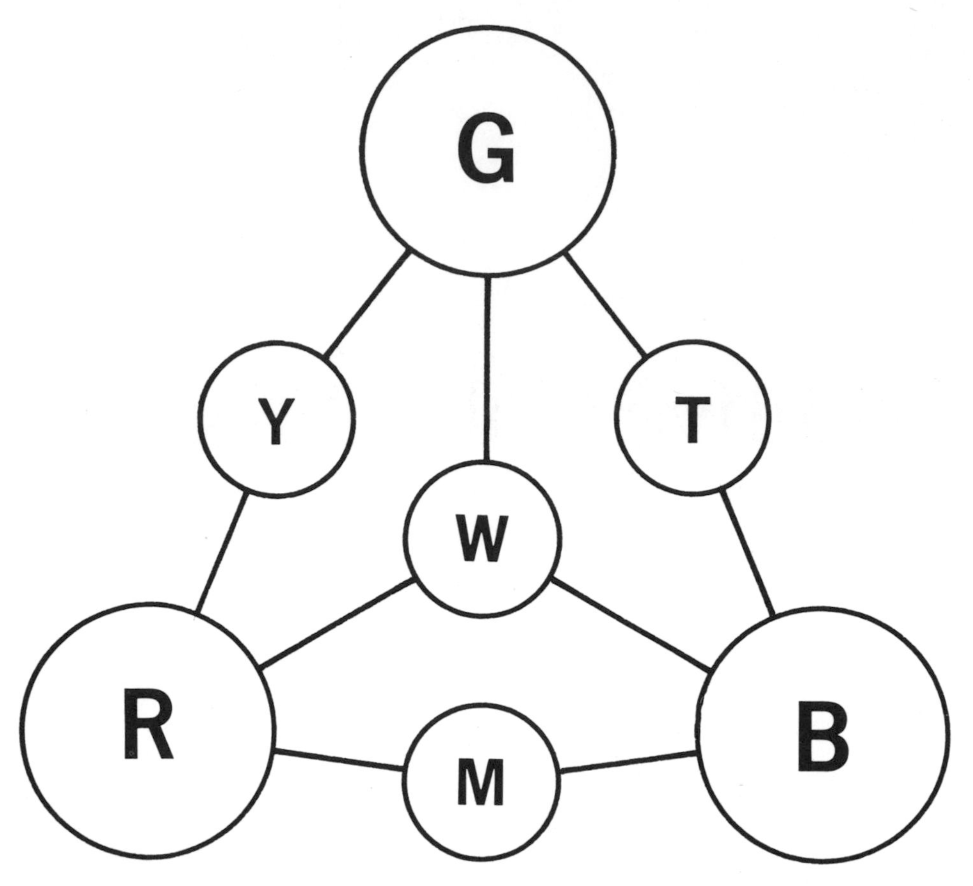

4-1. The additive primaries, in *light,* are red, green, and blue. In combination, they form magenta, yellow, and turquoise (cyan). All three additive primaries will form white.

orange paint, brown is formed. Sitting in a dark room, however, if an orange light bulb is dimmed with a rheostat, brown will not be seen. The eye will keep adjusting itself to the lower level of illumination and the orange light will merely grow dim. Related (surface) colors are light or dark, whereas unrelated (film) colors are bright or dim.

Additive mixture, encountered with colored lights, has three primaries: a red (ruby), green, and blue (cobalt). These three primaries will be seen when a spectrum is formed by directing a beam of white light through a prism. The red is warm in hue, the green is "grassy," the blue has some violet in it. See Figure 4-1.

Now, and as the reader may know, red and green light when combined will form yellow; red and blue light will form a magenta; blue and green will form a turquoise (peacock) blue.

Let it further be known that in process color printing filters in the three *additive* primaries are used by the photo-engraver to produce color plates that, in turn, will use *subtractive* printing ink primaries. The red filter will be used for the turquoise (cyan) plate; the blue filter will be used for the yellow plate; the green filter will be used for the magenta plate.

The additive primaries, red, green, blue, when mixed together with lights always add up to a *lighter* hue. All three additive primaries will form *white.*

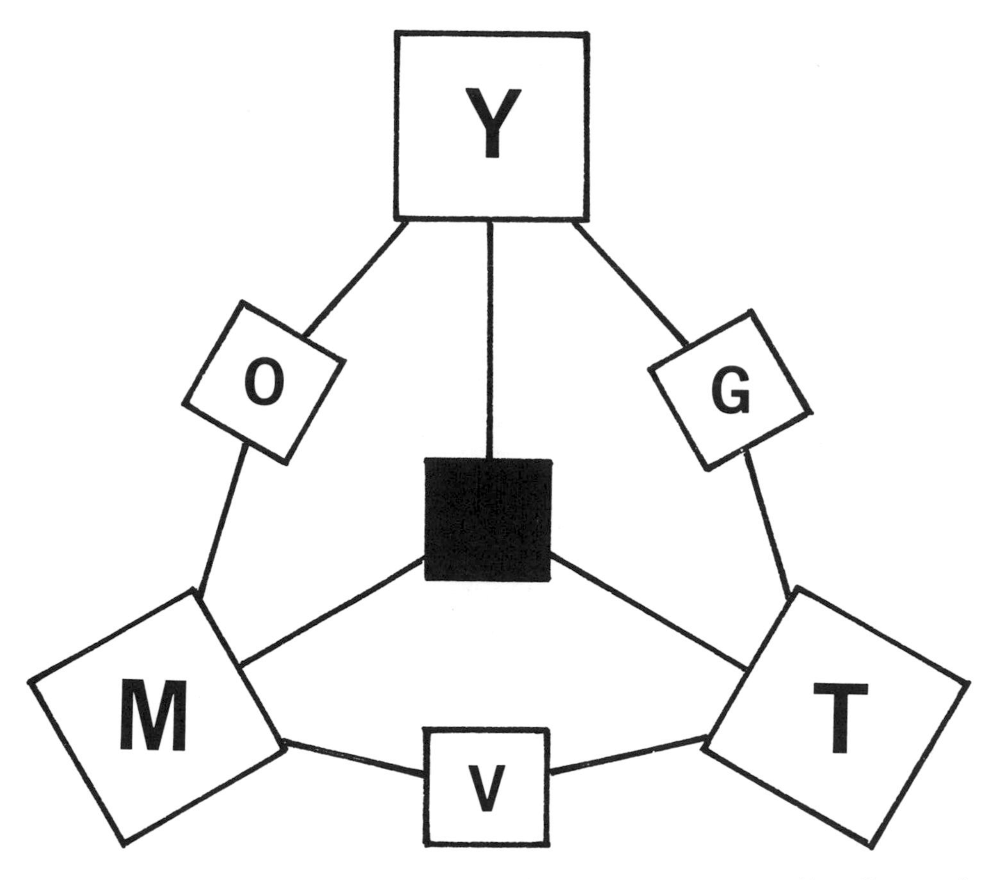

4-2. The subtractive primaries, in *pigments* and dyes, are magenta (red), yellow, and turquoise (cyan). In combination, they form orange, green, and violet. All three subtractive primaries will form black.

The subtractive primaries, magenta, yellow, turquoise (cyan), will tend to produce relatively darker colors. Magenta and yellow will form orange; red and blue will form purple; blue and yellow will form green. All three subtractive primaries will form *black*. See Figure 4-2.

Color television uses the additive primaries. Phosphors in red, green, and blue are electronically activated on the television screen to form other colors, as described above. The appearance of related colors in television (browns, blacks) is due to visual contrast over the *surface* of the screen. Here unrelated (film) colors are converted to the aspect of related (surface) colors.

Incidentally, most related (surface) colors can be changed to the appearance of unrelated (film) colors by viewing them through a small hole in a black card. The surface color, which would otherwise be localized, will, as seen through the hole in the card, seem to fill the space between the card and the colored surface itself.

To demonstrate facts about *additive* color mixture, the basis of much of the art of light and color, Rollo Gillespie Williams in his book *Lighting for Color and Form* presents two tabulations of the results of light mixtures—assuming that the three primaries are balanced to form white light.

RESULTANT COLOR	*PERCENTAGE OF FULL LIGHT*		*%*
	Red	*Green*	*Blue*
Blue	—	—	100%
Violet	8%	—	100
Cerise	50	—	100
Magenta	100	—	100
Crimson	100	—	66
Red	100	—	—
Orange	100	40	—
Yellow	100	80	—
Amber	100	100	—
Apple-Green	66	100	—
Green	—	100	—
Bluish-green	—	100	80
Peacock-blue	—	80	100
Turquoise	—	20	100

While pastel colors can be formed by the simple mixture of *white* light, many can also be formed with the primaries only, as Williams further charts:

RESULTANT COLOR	*PERCENTAGE OF FULL LIGHT*		*%*
	Red	*Green*	*Blue*
Lavender	33%	20%	100%
Mauve	66	20	100
Salmon Pink	100	80	100
Pale Green	33	100	100
Sea Green	8	80	100
Sky Blue	8	20	100
Pale Lemon	100	100	35

5 The Colortron

THE MAGIC OF COLORED LIGHT MIXTURES

The Colortron was born of necessity when I joined the Department of Design at the University of Kansas during the ninteen thirties. There I found that I had to teach an approach to color that was based on the mixture and relationships produced by color in light rather than by pigment. There was, however, no way of demonstrating the validity of this approach, and students had to take my word for it that red, green, and *blue—*not* red, yellow, and blue—are the primary colors in light from which all colors, including white, are produced by proper proportionate mixture; that the complement of red is blue-green, not green; et cetera.

I had to find a means to prove all this, and I recalled the device used by Paul Bornet in his art school in Paris. He had a large box, open at the front and fitted with concealed colored light bulbs, into which he would place paintings and other objects and, by turning on various bulbs, demonstrate color mixture and the effects of colored light on colored surfaces.

The Colortron, with the addition of rheostats and other refinements, was a further development of this idea. It not only solved the immediate teaching problem but proved to be an invaluable tool for experimentation and for demonstration of colored shadows, afterimage, simultaneous contrast, irradiation, color constancy, and many other phenomena of light and color that are present in nature but must be isolated in a controlled situation to be best observed—or to be observed at all. It also provides a quick and convenient way of exploring such matters as color harmony, color preference, color memory, the affective power of color, and color appropriateness in relation to form and function.

***Actually, the blue tends toward violet; but it is convenient, and perhaps less confusing, to simply say blue.**

5-1. Colortron closed to form carrying case.

5-2. Colortron showing control panel and background.

Several Colortrons have been built since the first experimental box, each an improvement in design and usefulness. Figure 5-1 shows the latest version: the prototype for a portable, self-contained model designed for quantity production for use in schools and colleges, where it could easily be carried from one classroom to another when wanted for demonstrations in the field of art, psychology, physics, photography, stagecraft, merchandising—wherever serious study of color is involved.

In Figure 5-1 the box is closed for carrying or for storage. It contains the remote control panel, the power cord for plugging into an electrical outlet, a number of charts and background panels, and several three-dimensional "props" such as the star seen in Figure 5-2.

Figure 5-2 shows the box opened, with the control panel and connecting cable out and the cut-out star in place in front of a white background panel. The power cord has been plugged in and the red, blue, and green bulbs have been turned on in proper proportion to produce a white star surrounded by colors on the background. The proper proportion is achieved through use of dimmers on the control panel, which will be described later.

Figure 5-3 is a view through the back from which the cover has been removed, revealing the row of light bulbs above and below in the coves behind the front. Each row consists of a white bulb, a daylight bulb, a red, a blue, a green, another daylight, and another white bulb—in that order. (Actually, a "white" bulb does not produce white light, although that term is used to designate the ordinary bulb used for illumination and will be so used in this text. It approaches white only when burned at the full prescribed voltage, rapidly becoming reddish as the voltage is reduced. For that reason, when low intensity is desired without change in color, a low wattage bulb must be used rather than decreased voltage. Similarly, a "daylight" bulb approximates the color of daylight only at full voltage, and a lower wattage bulb must be used when lower intensity "daylight" is desired.)

Figure 5-4 shows the Colortron in use as an illuminated display case for a chart. Here only the white bulbs are turned on. Other background panels make use of the white or daylight bulbs combined with colored bulbs to show changes in the appearance of objects when illuminated by light of various colors.

5-3. View of Colortron from rear showing arrangement of bulbs.

5-4. Colortron in use to display charts.

ADDITIVE COLOR MIXTURE

The circle of stars in Color Plate III shows the colors produced in the Colortron by the primary red, blue, and green bulbs, singly and in combination. The white star in the center is the product of all three together; and the secondaries—magenta, cyan (blue-green), and yellow—are combinations of pairs of primaries as indicated by the arrows. Around the white star in the center we see the complete spectrum on the background; all the colors of the rainbow combining to produce the white star.

The colors directly opposite each other across the circle are complements: that is, together they complete the requirements for producing white, the combination of all colors. Thus white will result from the combination of any primary and its opposite complement or secondary; also from the three secondaries—magenta, cyan, and yellow—since they are composed of the primaries. (Combined secondaries will also produce other colors; but that gets into subtractive mixture, and we are here concerned only with additive. In the Colortron we use only the additive primaries red, blue, and green since with these we can get white and all other colors in maximum saturation.)

(The color effects shown here are reproduced directly from Ektachrome transparencies of the actual subject matter, as is true of all the color plates.)

COLORED SHADOWS

Probably few people ever see color in shadows, although it abounds in nature and in all our environment. For example, shadows among tree-covered hills take on a purplish cast, as do shadows in the foliage of the trees themselves; and shadows in snow under neon lights are bluish green. But because color effects all around us are so diffused, and because of our perceptual tendency to color constancy, these shadow colors go unnoticed by the untrained eye. It is only under controlled conditions such as are provided by the Colortron that these effects are strikingly evident. This phenomenon has long been known to artists and philosophers—notably Goethe who, in his *Theory of Colors,* describes his extensive experiments in the field. The tendency of shadow colors is always complementary to the color of the shadow-casting light, as shown in Color Plate IV. Here the shadow colors are unusually pronounced because the background was a light neutral gray without any color of its own to interfere, and because in each of the three examples the shadow-casting light was of a single color. The only other illumination was from a very small "daylight" bulb placed at a distance from the Colortron so that its light fell only upon the background. Without this illumination the shadow colors could not be distinguished. (Goethe points out the need for this neutral illumination of the background in order to make the shadow colors discernible.)

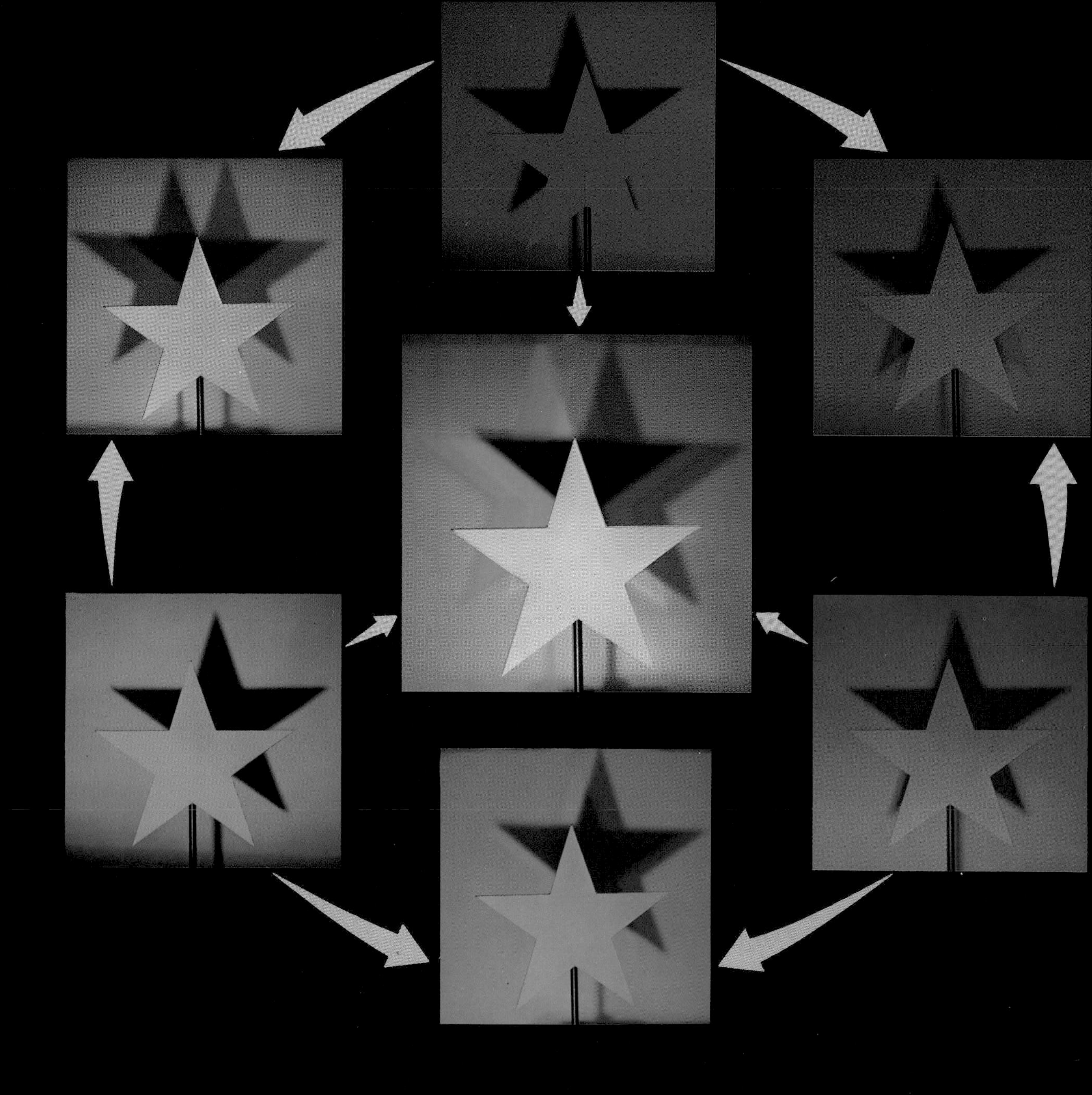

PLATE III

The additive color circle. Note how it differs from the subtractive pigment circle to which we are accustomed. Here the primary colors are red, blue, and green, and the secondaries are cyan (blue-green), yellow, and magenta. All the primaries (or all the secondaries) together produce white, as shown in the center star.

In *A* of Color Plate IV the background was a very light gray instead of white in order that the star would stand out clearly, and the star shape was used for casting the shadows because its many sides and angles provide maximum possibility for observing color contrasts, as demonstrated in the central panel.

The Ektachrome transparencies of the colored shadows seemed too good to be true and, wondering whether there might be some unsuspected factor responsible for the apparent success of the experiment, I tried other kinds of film, including Polaroid. The results were the same—which would seem to refute the long-held belief that such shadow colors are subjective and to indicate that they actually exist.

BASIC CHARTS AND BACKGROUNDS

One use for the Colortron in the classroom is the effective, properly lighted display of pictures, charts, and teaching aids. Color plates IV and V show charts that are basic to demonstrations of color phenomena. These are most easily and attractively made by using colored paper on mounting board cut to fit the opening in the back of the Colortron.

B of Color Plate IV shows the "light circle" and a "pigment circle" side by side and provides a means of pointing out the differences between the two, and of dispelling, as far as possible, the confusion that is bound to persist in a discussion of additive color mixture. (Also see Chapter 4.)

C of Color Plate IV introduces the mystery of negative afterimage and induced color. If we gaze for ten or more seconds at the black dot surrounded by circles of color and then quickly shift our eyes to the lone black dot at the right, we should see around it circles of colors complementary to those around the first dot. Some viewers will not be able to see the complementary colors on the first trial: a few will see them after fixating on the first black dot for only a second. This experiment carries over into the observation of afterimages in many situations.

D of Color Plate IV presents another manifestation of induced color: simultaneous contrast, or the apparent change in a color caused by others adjacent to it or surrounding it. It also shows the apparent change in size of an area surrounded by a lighter or darker area because of the effect of irradiation.

At the top are two rectangles of exactly the same size, cut from the same sheet of gray paper. The one at the right, surrounded by white, seems smaller and darker than that surrounded by black at the left.

At the bottom are two identical green rectangles. The one on the right seems smaller than the one on the left because it is seen against a lighter background. The green rectangle on the right seems bluish because it partakes of the complementary color of the surrounding yellow. The green at the left seems more yellowish because it is invaded by the complement of the surrounding blue.

In the middle pair of identical gray rectangles the one on the right seems warmer because it is surrounded by green, while that on the left seems cooler because of its red background.

Various panels can be made, depending on the subject under investigation. One, for example, could deal with legibility of printed matter in different color combinations and under different levels of illumination. Another could show apparent differences in size and weight of identical objects when painted different colors. Another could show how some colors seem to advance while others retreat.

A panel made of bold areas of colored paper is shown in *A* of Color Plate V. At the upper left the panel is illuminated by red, blue, and green bulbs. In the actual full-size chart a striking example of simultaneous contrast is observable. If the gaze is focused on the juncture of two adjacent colors each will seem brighter along the edge where it meets the other color and will seem grayed or somewhat neutralized farther back from the edge. This is particularly true when the two colors are complementary or nearly so.

At the upper right the panel is illuminated by red bulbs only; at lower left by only the blue; at lower right by only the green—showing the effects of different colored lights on colored surfaces.

B of Color Plate V shows a panel covered with a piece of actual textile. At the left it is illuminated by the white bulbs. At the right we are using only the blue. Notice how the other colors disappear, showing that a color is visible only when the illuminating light contains that color.

A of Color Plate VI shows what happens to a girl's face under light of different colors. Almost, that is, because these are color photographs of a plaster mannequin kept on hand for demonstrations, since a pretty girl is not always available. When one is available to stand behind the Colortron with her head in the opening, however, it not only adds much zest to a demonstration but shows the changing effects much more convincingly. By adjustment of the dimmers on the control panel any desired color effects can be produced to illuminate the face; and experiment can determine what make-up and what color of dress are most attractive in a given lighting situation. A flattering light is pink, or violet along with pink—a combination possible on the stage through the use of such filters as "Shubert pink" in the spotlights. While this is hardly practicable at the average school party or in the home, a discreet combination of pink or violet bulbs and white bulbs may bring the desired results.

In *B* of Color Plate VI a painting has been placed behind the opening in the back of the Colortron. At the left it has been lighted by only the white bulbs. At the right the white bulbs have been turned off and the red, blue, and green turned on in correct balance to show the scene as it would appear in daylight, as the artist saw it when he was painting it. This method of illuminating paintings—by multiple sources of colored light—while beyond the resources of the average home, recommends itself to

A

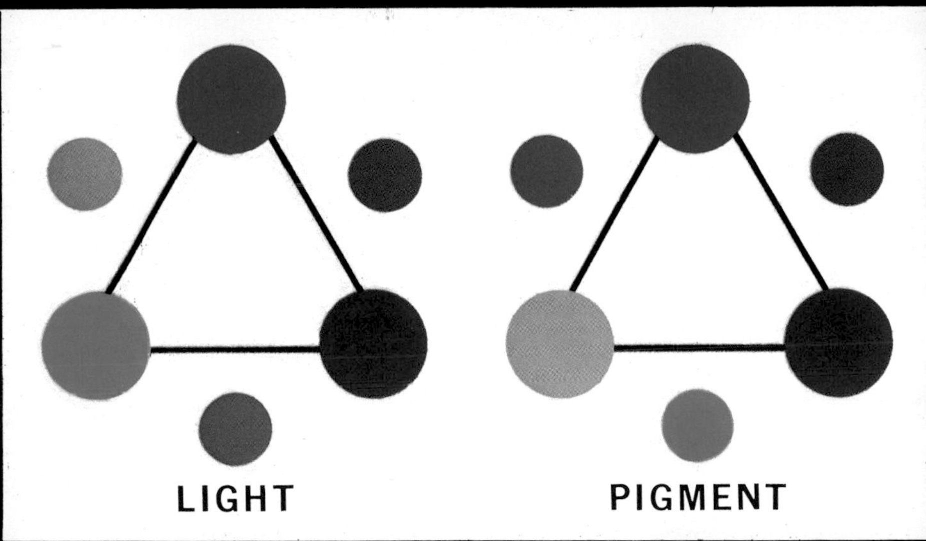

B

(*A*) The colors of shadows are the complements of the chromatic illumination. (*B*) The color circle of light compared with that of pigment. (*C*) Negative after-image. After gazing at the black spot at left, if we quickly shift our eyes to the one on the right the effect will be apparent. (*D*) Simultaneous contrast, showing the effect of the surrounding areas on the color and apparent size of the inner rectangles.

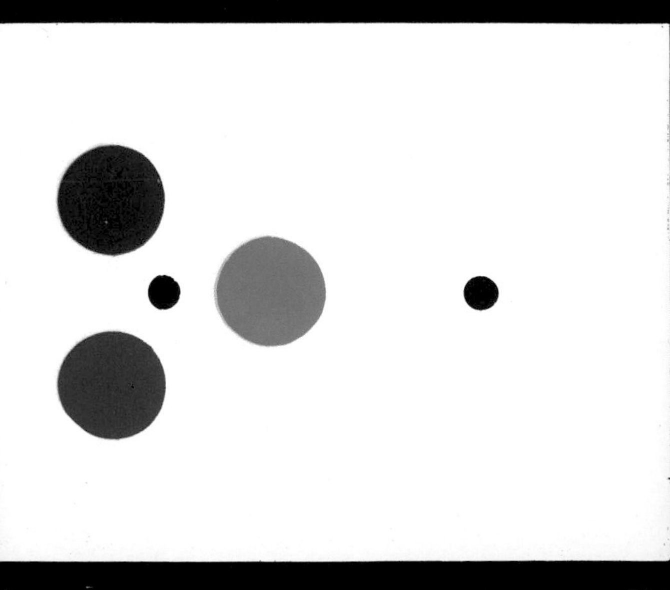

C

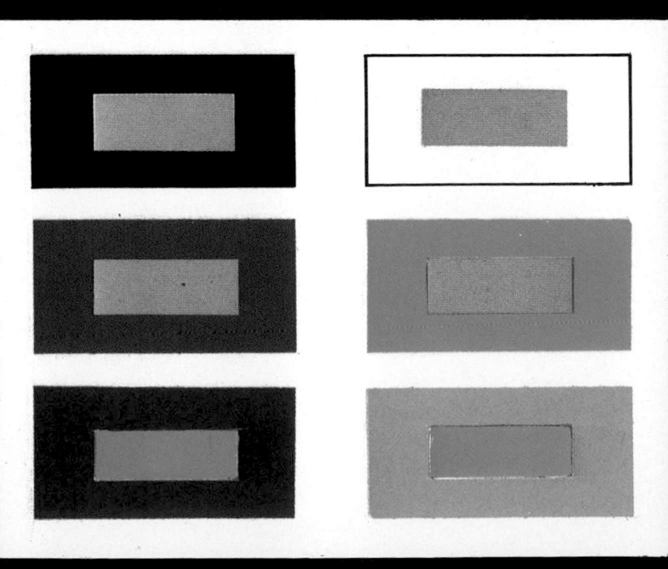

D

PLATE IV

A

Effect of chromatic light on colored surfaces. (*A*) A pattern of colored papers under white, red, blue, and green light. (*B*) *Left:* A textile under white light; *right*, under blue light.

galleries. As Williams explains in his *Lighting for Color and Form,* the colors need not be restricted to red, blue, and green but, in the interest of economy and greater efficiency, can include white; or can be the secondary colors magenta, peacock-blue (cyan), and yellow.

C in Color Plate VI shows the importance of proper lighting in merchandise display. A bottle of ketchup was, by chance, the object used in the first demonstration on this subject, and the reaction was so favorable that no other object has been used since—despite the ketchup bottle's utter lack of aesthetic appeal. At left the ketchup is illuminated by red light only, giving an anemic, washed-out effect. Next, the result is repulsive when only green light is used. Next, ordinary white bulbs show the ketchup as we normally see it. Last, at right, we add glamor by using red light along with the white. We could add still more glamor by using red, blue, and green light in proper balance, but few stores are equipped for such lighting, and the combination of white and red bulbs does very well.

This is a sure-fire demonstration: the intended effects under the various lighting conditions are immediately obvious, and the effect produced by the green light always brings audible gasps of revulsion. This leads to a discussion of color appropriateness in packaging and in other fields, and the realization that the odious green of the ketchup could be acceptable and pleasing in another context.

Since the color of food is so important, another demonstration, *A* in Color Plate VII, is offered to show the effects of colored light on fruit. At upper left the fruit is illuminated by a white bulb: at upper right the lighting is by red, blue, and green bulbs in proper balance to enhance the color of the fruit. Below, the fruit is shown as it appears when lighted in turn by a red, a blue, and a green bulb.

B in Color Plate VII shows how two scenes, painted in different colors, can be shown individually by illuminating them by light of their respective hues: another example of the effect of colored light on colored objects. Above is a panel on which a city scene is sketched in red and a rural scene in blue, shown under ordinary light. At lower left the panel is shown under red light which causes the city scene to disappear because it cannot be distinguished from the over-all red on the background. The blue rural scene, however, comes through very strongly because the red light causes the blue drawing to appear black. At lower right the panel is shown under blue light, and the blue drawing disappears because it cannot be distinguished from the blue on the background. The red city scene, however, is very obvious because the red lines seem black under the blue light.

This effect is sometimes used on the stage for a quick change of background, particularly when the scenes are symbolic rather than naturalistic. It is essential that the two scenes be painted under the same colored light under which they will later be shown; and dyes instead of paints should be used because the latter may be discernible from the background because of their difference in texture, regardless of the color of the light.

The same principle is sometimes used in children's books to show two different illustrations printed in the same area. In this case, instead of being seen under different colored lights, the picture area is viewed through different colored gelatin filters that are provided in the book.

In *A* of Color Plate VIII is another example of change of scene through change of color of light. In one instance a painting of an outdoor scene in normal daylight is illuminated by white bulbs. Under red light, however, we have a snow scene—all the red areas of the painting being washed out and appearing white. In addition, both a man standing in the center and an awning on the house disappear. Careful study of the action of red light on the original pigment colors in the painting is responsible for this unusual legerdemain.

While the didactic and investigative resources of the Colortron were being explored, the aesthetic possibilities of the color effects became apparent. In *B* of Color Plate VIII the play of colored light on abstract cardboard forms and the changing shadow patterns on the background as the forms were moved about by hand suggested the use of a small motor-driven turntable to rotate the forms slowly and create moving compositions in color. Then came the idea of putting a tracing paper screen across the opening in the rear and observing the effects from behind.

Thus the Colortron became a creative instrument, the progenitor of the Sculptachrome and the Chromaton, both of which will be described and illustrated in later chapters.

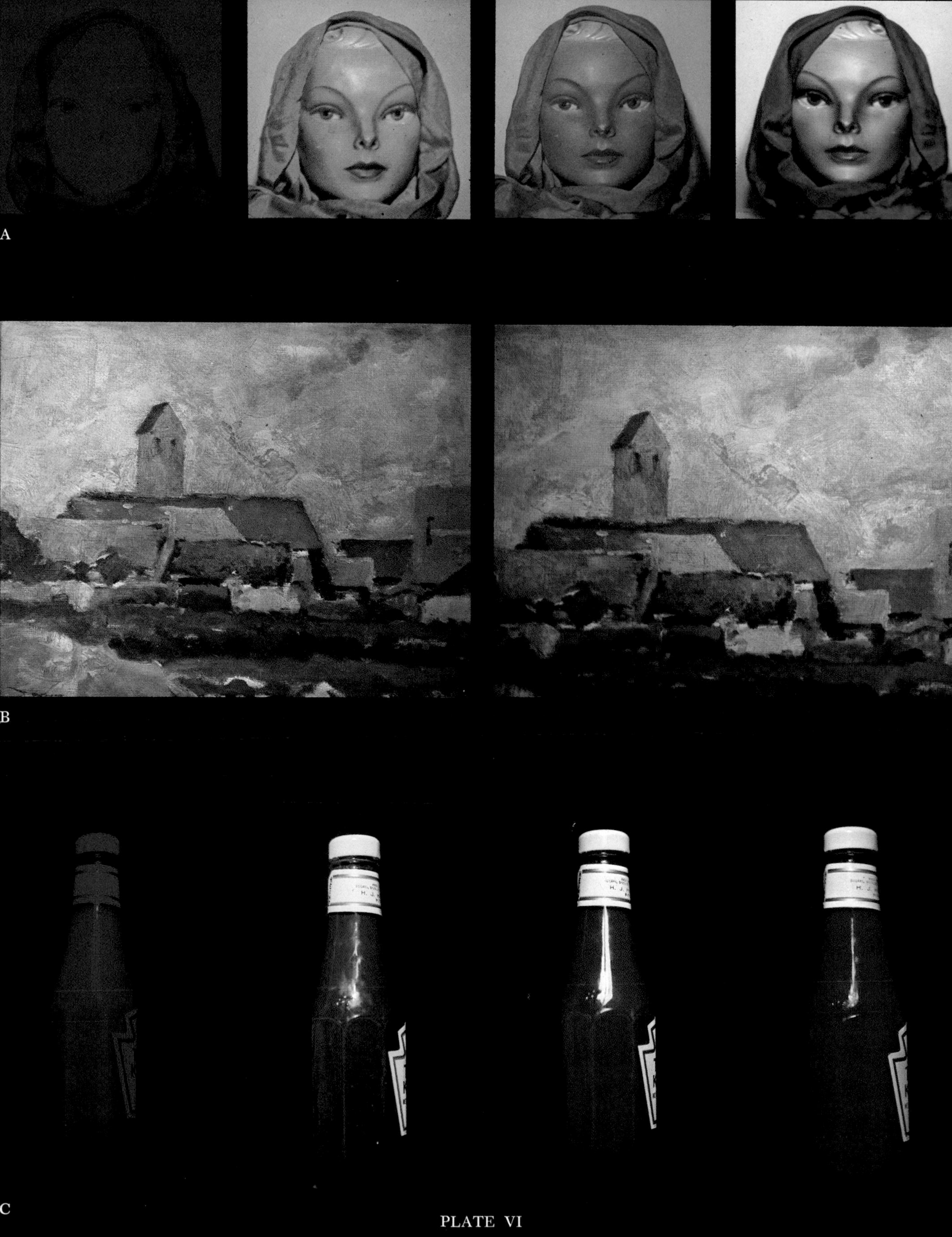

PLATE VI

A

Opposite:
(*A*) A girl's face under red light, green light, blue light, and under all three combined. (*B*) *Left*: A painting under ordinary incandescent light; *right*, under a balanced combination of red, blue, and green light. (*C*) *Left to right*, A bottle of ketchup under red light, green light, white light, and white light combined with red light.

(*A*) *Top left:* Fruit as it appears under ordinary incandescent light; *right*, under a balanced combination of red, blue, and green light; *below*, under red light, blue light, and green light. (*B*) *Center:* Superimposed scenes, one in red and one in blue, are shown under white light; *below left*, the effect under red light; *right*, under blue light.

B

PLATE VII

HOW TO BUILD A COLORTRON

Since a Colortron cannot be bought, it will have to be built; and following are drawings and descriptions for constructing one at minimum cost. The undertaking presupposes the availability of a school shop and the active cooperation of an instructor and two or more versatile students who, as a class project, will do the assembling, painting, simple wiring of the instrument, and the preparation of the charts and backgrounds; also, an appropriation of a hundred and fifty dollars.

As previously shown in halftone illustrations (Figures 5-1, -2, -3, -4) the Colortron is basically a twenty-four-inch cubical box (Figure 5-5) from which a part of the front, sides, and back are cut away, fitted with colored light sources and means of controlling them.

Figure 5-6 is another view of the Colortron cabinet from the rear toward the front. Figure 5-7 shows a straight-on view of the bulb arrangement and a cross section of the bulb and reflector assembly in the lower cove in the front of the cabinet. An identical assembly is in the upper cove.

5-6. Detail of Colortron from back to front.

A 16″ x 20″ opening in the rear of the cabinet is fitted with a channel at top and bottom for the insertion of charts and backgrounds. A "stage," 3″ x 6″, flush with the lower edge of the opening, is for the display of three-dimensional objects under various conditions of illumination.

The upper and lower coves at the front are provided with hinged lids to give easy access to the bulbs. The lid of the upper cove is held closed by hooks that engage pins, or by simple latches at either side of the cabinet. Across the inside of the coves and behind the bulbs are tin or aluminum reflectors pierced by holes through which the bulbs are inserted into

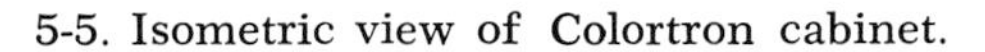

5-5. Isometric view of Colortron cabinet.

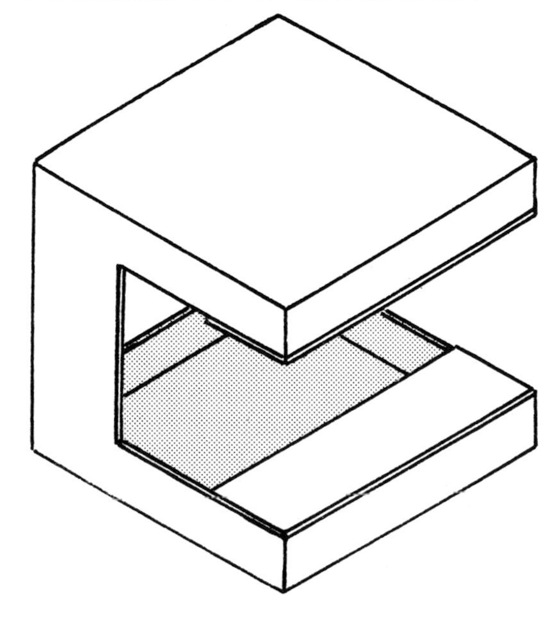

5-7. Bulb arrangement for Colortron.

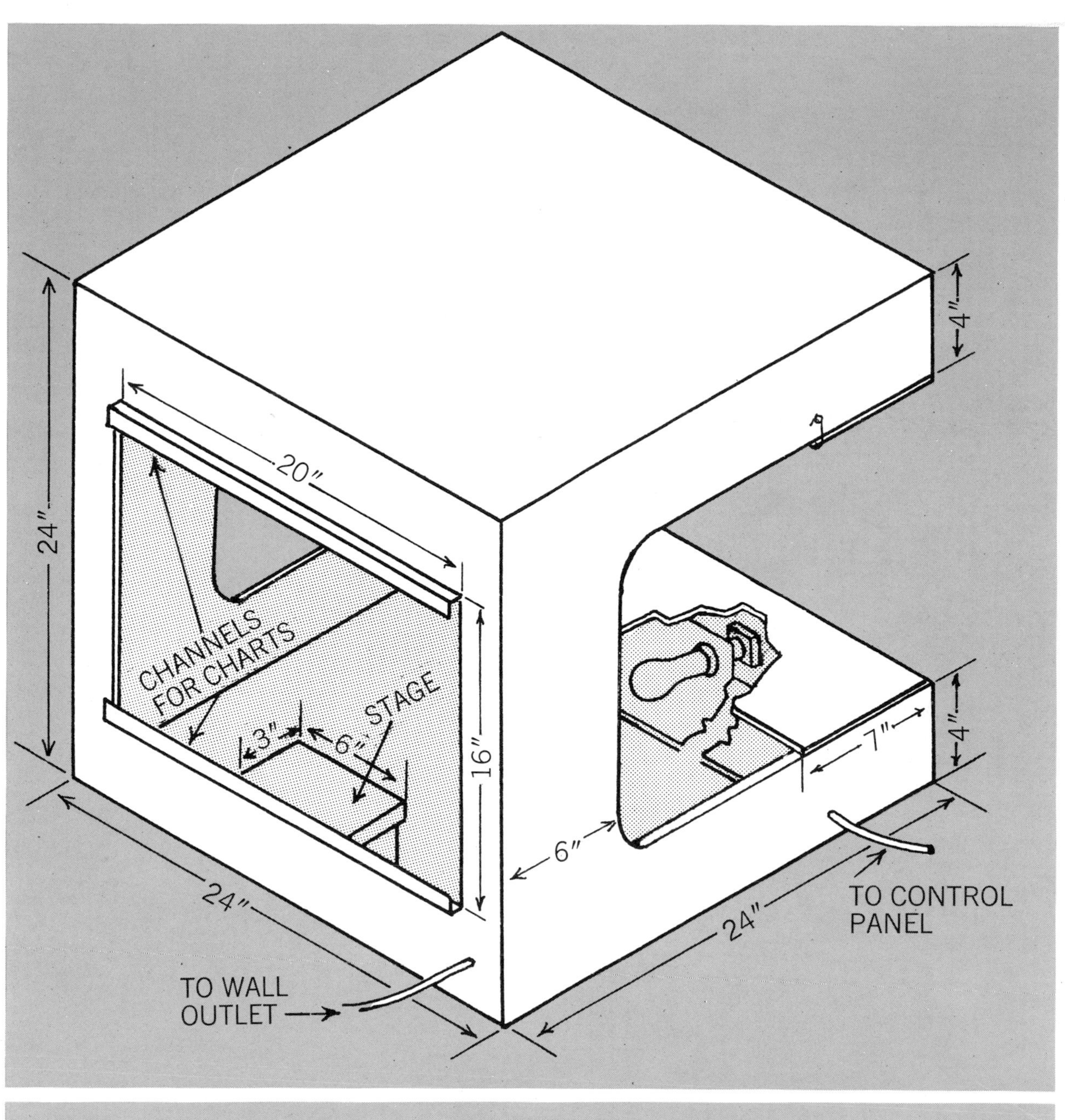
24"
20"
CHANNELS FOR CHARTS
STAGE
3"
6"
16"
6"
7"
4"
4"
24"
24"
TO WALL OUTLET
TO CONTROL PANEL

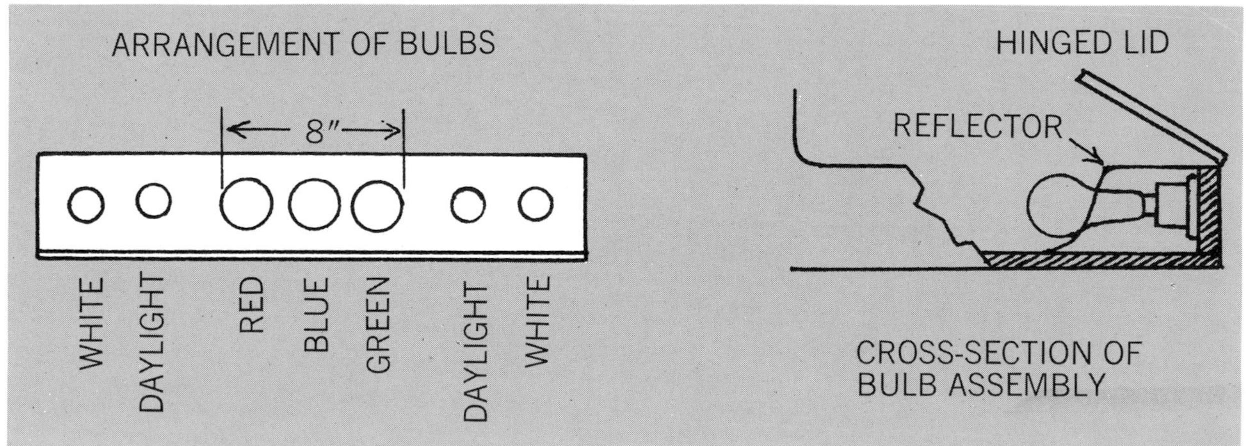
ARRANGEMENT OF BULBS
8"
WHITE
DAYLIGHT
RED
BLUE
GREEN
DAYLIGHT
WHITE
HINGED LID
REFLECTOR
CROSS-SECTION OF BULB ASSEMBLY

A

B

C

the sockets. (See cut-away portion in Figure 5-5 and cross section at right in Figure 5-7.)

The arrangement of bulbs (at left in the lower panel) is identical in the upper and lower cove except that the relative positions of the red and the green bulbs should be reversed so that the upper red bulb is directly above the lower green and the upper green is above the lower red. This provides the best possible blending of the colors of light falling on the background panels and on objects placed on the stage. To the same end it is important that the three colored bulbs in each cove be within a span of eight inches.

The 10-watt white and daylight bulbs, the sockets and the necessary wire should be available at any electrical supply store but the colored bulbs may present a problem. They must be of the type known as "natural colored." That is, the color must be in the glass itself, and the filament must be visible when the bulb is turned on. These are made by Sylvania and by General Electric and come in a limited number of colors and sizes. The bulbs needed for the Colortron are 60 watt primary red, 100 watt primary blue and 60 watt primary green, and several of each should be kept on hand.

Natural colored bulbs were once widely used in stage and display lighting but have been to a great extent replaced by plain clear bulbs in combination with filters. These bulbs have been used in the Colortron because they are the simplest and most convenient means of obtaining the modest quantities of colored light required. However, they may be in such scarcity that clear bulbs and filters will have to be used. This change will not affect the design of the Colortron but will require construction of enclosures for the bulbs and their filters. The placement of the bulbs will be the same.

Figure 5-8 shows the filter housing installed over the clear 100-watt bulbs that have taken the place of the colored ones. The central part of the cove covering has been cut away to make room for the housing, leaving a cove on either side for the white and laylight bulbs. The filter on the left is red, that in the middle is blue, and that on the right is green—corresponding to the arrangement of colored bulbs shown in Figure 5-7. An identical filter housing is installed in the upper cove.

Figure 5-9, on a larger scale, shows another view of the filter housing, looking into the three compartments that fit over the bulbs, and showing the flange that fits over the front wall of the cove to which it will be fastened with screws.

For the filters, a set of six 8″ x 10″ cellulose acetate sheets is available from the Edmund Scientific Company of Barrington, New Jersey. The set contains both primary and secondary colors, and while the primaries are basic to the repertoire of the Colortron, the secondaries are useful, too, for experimentation.

The sheets can be cut to fit into the slots of the filter housings, and two or more filters can be used for each color to obtain the desired density and hue. For example, a single sheet of blue may not be dark enough and not quite the right color to combine with the red and green to produce white

Plate VIII. (A) Top left: A painting seen under ordinary light; *below,* the painting as it appears under red light only. *(B) Top right:* First steps in the development of the Sculptachrome, and the Chromaton, *below. (C)* A simple demonstration of the effectiveness of colored shadows cast by red, green, and blue lights.

when all three colors are turned on to illuminate an object in the Colortron. A second thickness of blue and one of magenta can be superimposed on the single blue sheet to give it the saturation and the tinge of violet that it should have.

Minute construction details of the Colortron cabinet are not shown as these are entrusted to the ingenuity and judgment of the builders, and may be dependent on the resources at hand. However, the dimensions specified have been determined by careful study and experiment and should be strictly adherred to.

The best way to build the cabinet is to cut the two sides out of ¾″ plywood and fasten to these the top, bottom, back, and the upper and lower coves. The interior must be painted flat black or, preferably, lined with black velour or flock paper (also available from the Edmund Scientific Company).

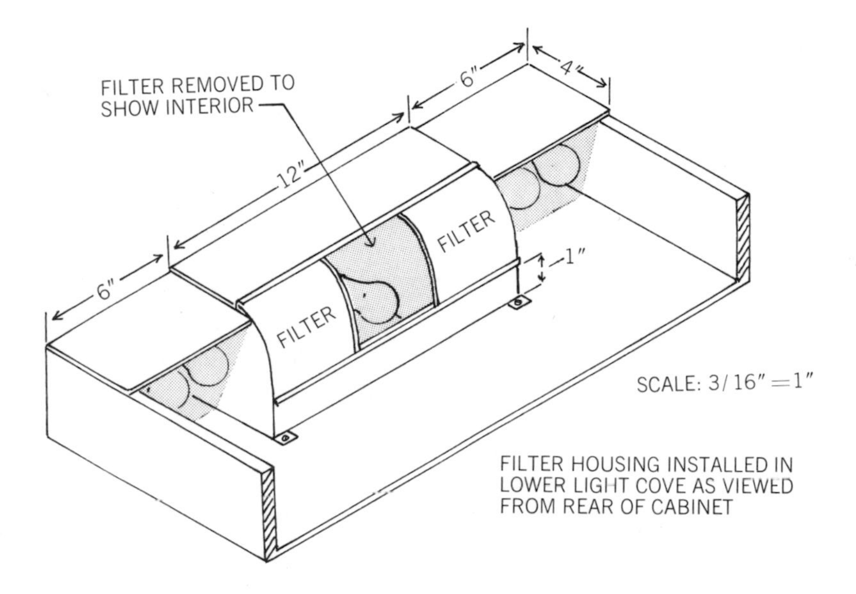

5-8. View of filter housing in Colortron.

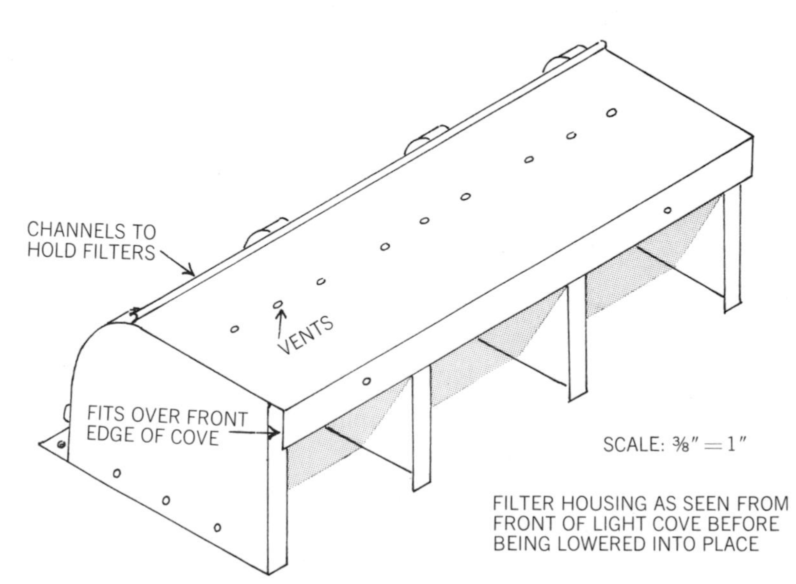

5-9. Enlarged view of Colortron filter housing.

The remote control panel (Figure 5-10) enables the demonstrator or experimenter to observe the effects produced in the Colortron from a vantage point in front of the instrument. It contains switches for the white and daylight bulbs; also, master switches for all the color, all the white, all the daylight circuits, and a main switch for the panel as a whole.

The dimmers are of the solid-state type used to replace wall switches in homes and offices. They are available at most hardware and electrical supply stores and from the Edmund Scientific Company.

Toggle switches are recommended because they enable the operator to see at a glance which ones are in closed position before the master switch or main switch is turned on, thus providing for pre-set combinations. For convenience in manipulation, the switches should be oriented so that the handles move in horizontal direction.

In the drawing the dimmers and switches for the upper colored bulbs are shown in the upper half of the panel and those for the lower red, blue, and green are in the lower half. The color master switch, which turns all the colors on or off at once, is between the upper red and blue switches. The main switch, which controls the whole panel, is between the lower red and blue. At the upper right of the panel are the switches and master switch for the four white bulbs. At lower right are the four switches and the master for the daylight bulbs. On the side of the panel box, at lower right, is the outlet for the connecting cable.

The dimmer knobs should be provided with calibrated dials, as indicated on the upper red dimmer, so that dial settings required for various effects can be recorded and reproduced at will—eliminating the need for fumbling and experimenting when a specific effect is to be demonstrated. The dials can be made of white paper pasted to the panel beneath the knobs, and the calibration from 0 to 100 will be determined by the particular brand of dimmer used. Probably not more than half the circumference of the dial will be used in progressing from off to full on.

Dimensions for the panel and box are not given on the drawing, since much depends on the material (preferably aluminum) of which the panel box is made, the material (preferably 3⁄16″ bakelite) used for the panel, the dimensions of the brand of dimmers used, and the choice of the user of the panel, who may prefer a more or less generous spacing of the controls. For a panel of convenient size for holding in the hands while demonstrating, and with adequate spacing between the components behind the panel, the dimensions should be about 8″ x 12″ x 2½″.

If desired the panel can include a small pilot light to aid in finding the controls in the dark, but this would add to the size, and with practice the user will learn to find the controls by feel.

The cable connecting the control panel to the Colortron cabinet can be made up of 15 eight-foot lengths of #18 insulated wire, bound together and encased in a spiral winding of insulating tape. The cable can be longer or shorter, but the length of eight feet has been found satisfactory.

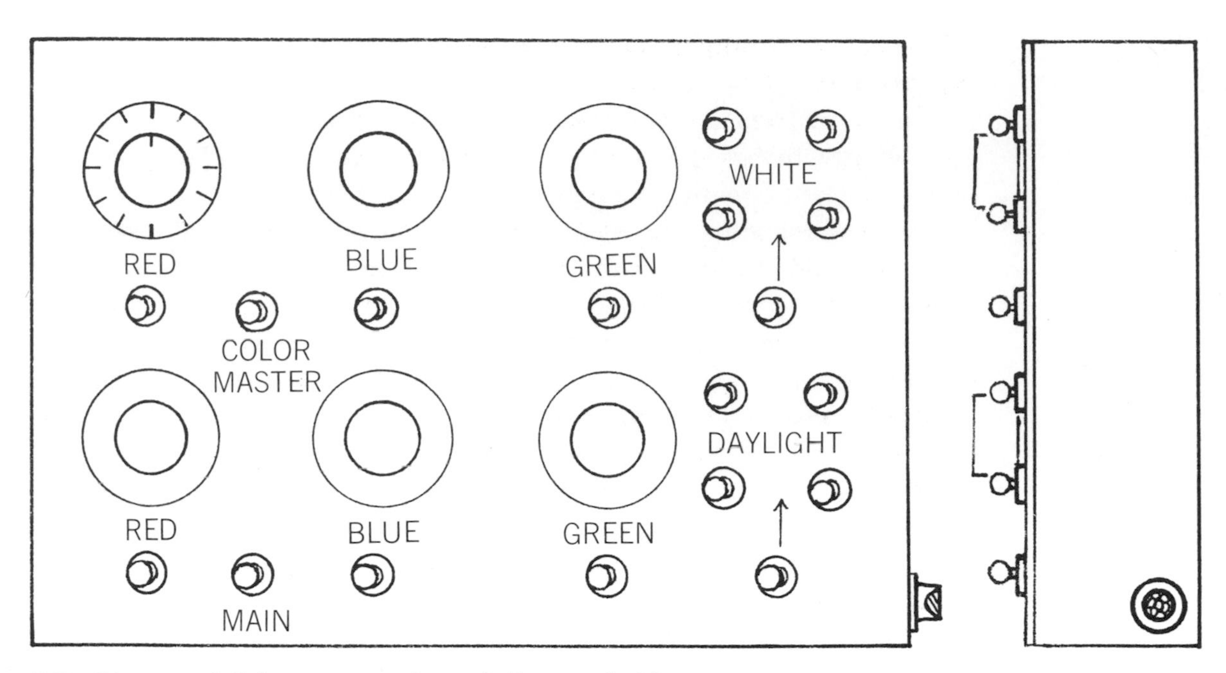

5-10. Diagram of Colortron control panel, front and side.

Figure 5-11 is a wiring diagram which should be self-explanatory. Fuses are indicated in the colored light circuits for protection of the dimmers in case of short circuit, which sometimes happens when an incandescent filament burns out. The small 1-ampere cartridge fuses are held in fuse clips which are mounted inside the coves of the Colortron cabinet.

5-11. Wiring diagram for Colortron.

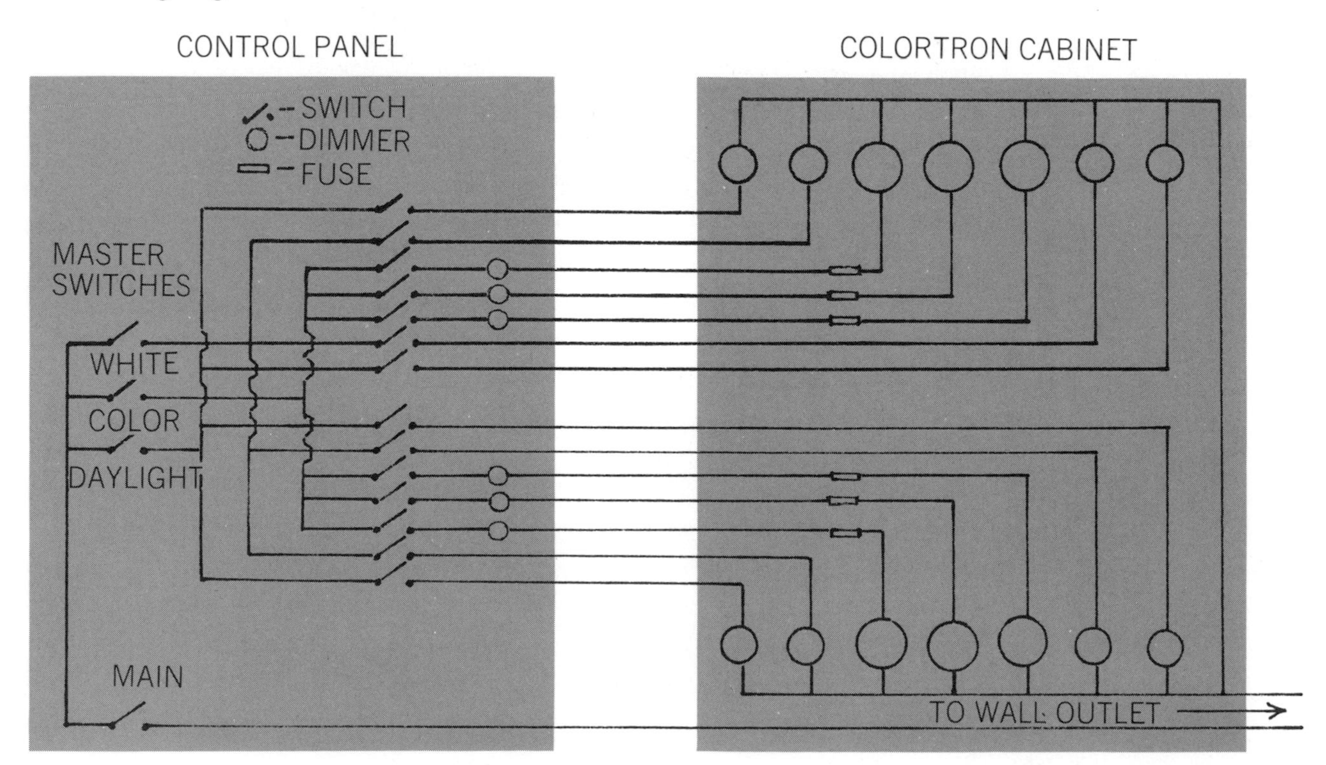

There may be times when it is desired to use the remote control panel to control a lighting situation outside the confines of the Colortron. This can be accomplished by removing the color bulbs from their sockets in the Colortron and replacing them with extension cords leading to the external light units.

The Colortron described is intended for small groups and is not adequate for large audiences. However, the principles and effects can be shown on larger scale by constructing larger props, backgrounds, and charts, and using larger light sources and controls such as are used in theatrical productions.

To end this chapter, two simple demonstrations are here described and illustrated, both of which will feature the beauty of additive colors and which will be easy and economical to make.

The first, which I call the Spectrum Footlight, uses three spotlights (red, green, blue) mounted close together on a board (Figure 5-12) and placed at floor level near a white or pale gray wall. The three spots are directed upward toward one area on the wall, and the combination of their colors will approximate white. However, when any action is introduced between the spots and the wall—students, teachers—shadows will be surrounded by a spectrum of bright red, yellow, blue, as in *C* of Color Plate VIII. The effect is startling and beautiful, and will invite student participation. Suitable lamps for the purpose are General Electric 150-watt Dichro Projector Spots in red, blue, and green, or Sylvania 150-watt Glass-Spots, Red-Stained 6R, Blue-Stained 6B, and Green-Stained 6G.

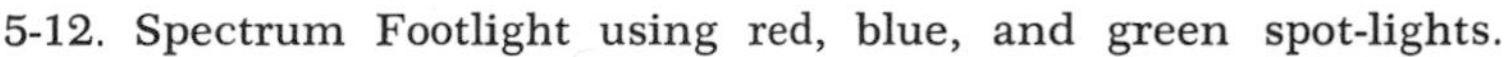
5-12. Spectrum Footlight using red, blue, and green spot-lights.

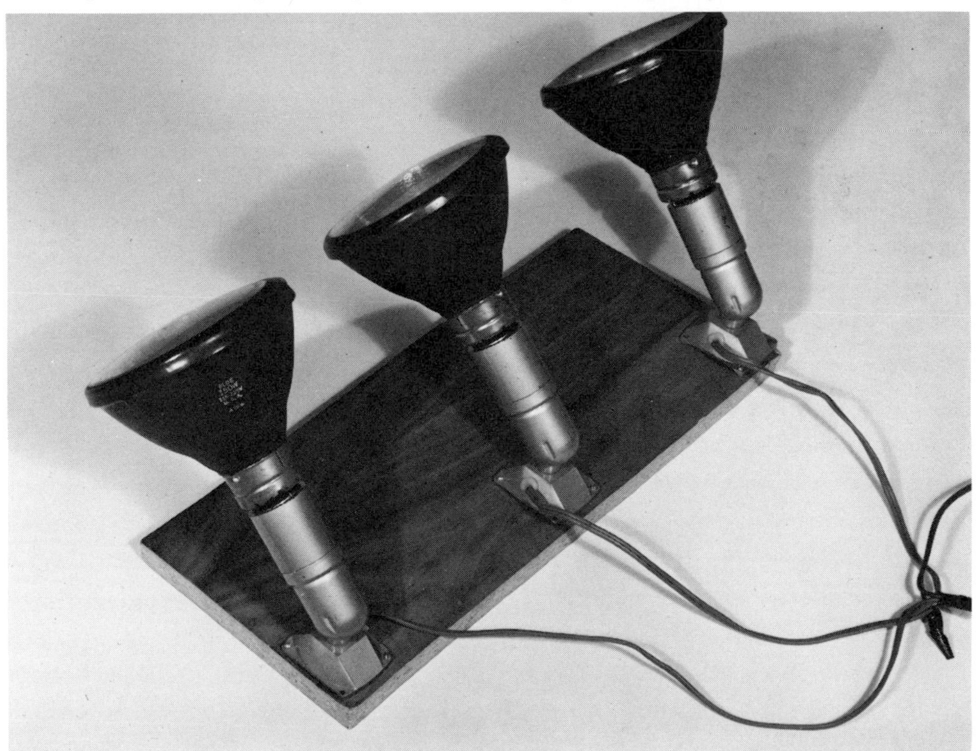

5-13. View of Light Cove showing control panel and arrangement of bulbs.

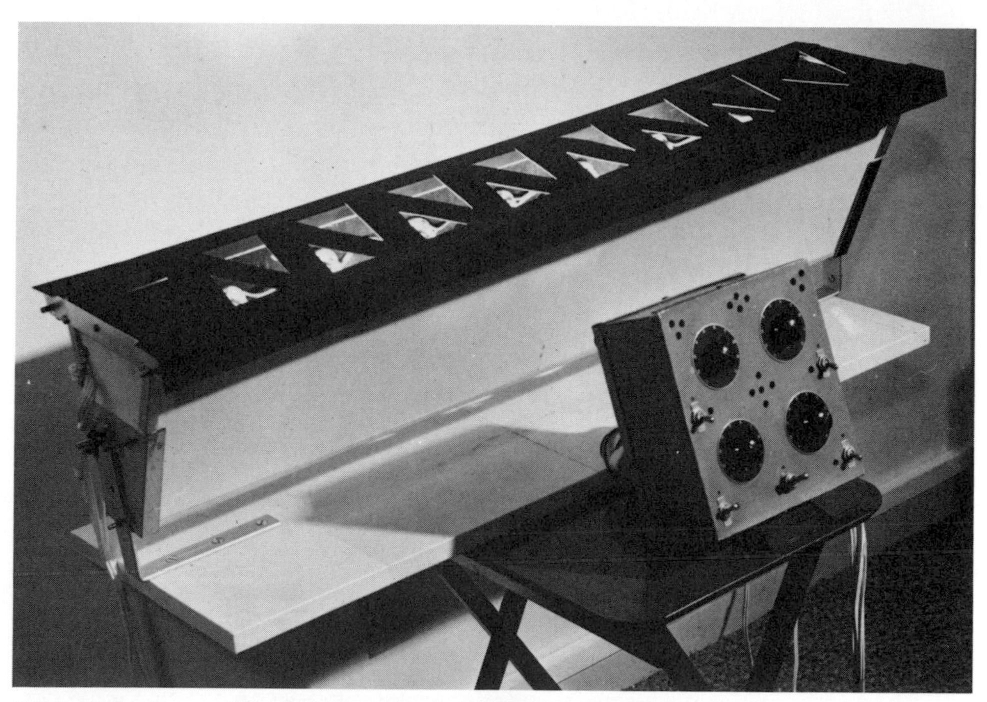

5-14. Light Cove with stationary mask in place.

The second demonstration is composed of a cove in the shape of a flower box. A variation of this was designed years ago by M. Luckiesh, an eminent lighting engineer. The cove is painted black inside and equipped with a row of sockets (Figure 5-13). These sockets hold ordinary transparent-color sign lamps (such as G. E. Lexan) in red, green, blue, white, and are controlled by switches or low cost solid-state dimmers. Cutout masks or stencils in geometric patterns are laid on top of the cove in order to cast shadows when the bulbs are lit (Figure 5-14). With the cove placed high on a wall, the wall area above it and the ceiling will be illuminated by a blaze of spectrum hues, and these pure colors will shift and move as the dimmers are manipulated. The cove can be as narrow or as wide as desired.

A motorized version is shown in Figures 5-15 and 5-16, in which a continuous stencil is moved across the top of the cove. Figure 5-16 shows the stencil above the cove before it is put over the latter and made into a continuous belt. Color Plate IX shows two examples of the effects produced on a wall.

5-15. Light Cove in use.

5-16. Motorized stencil belt to add movement in use of Light Cove.

6 The Sculptachrome

A NEW APPROACH TO DESIGN

From its lowly beginning as a side effect in the Colortron, the Sculptachrome emerged as an independent instrument of creative expression; a combining of kinetic sculpture and mobile colored light, with dramatic background shadows of flowing colors, as in Color Plate X.

The Sculptachrome is very simple in principle (Figure 6-1): three colored light bulbs (red, blue, and green) with a dimmer for each; a slow-moving turntable bearing an abstract construction of cardboard or of balsa wood painted white—all contained in a base or box about a foot wide.

From here on I shall, as far as possible, deal only with basic elements and principles without giving minute dimensions or plans for construction, it being assumed that professional help will be used in technical problems. It is further assumed that anyone interested in these projects will want to develop them in his own way for his own particular purpose. For example, one teacher may want a Chromaton for permanent classroom installation, while another, who is interested in extramural lecturing, may want a collapsible, portable unit that will fit into the trunk of his car. (I have built both.)

6-1. View of Sculptachrome with abstract construction.

PLATE IX

6-2. Rear view of Sculptachrome showing arrangement of bulbs and turntable.

The turntable is a 4″ disc fixed to the shaft of a 1-rpm synchronous motor (obtainable from Edmund Scientific Company). Slower motors are available, but this speed seemed about right for the purpose—classroom demonstrations of changing mixtures of colored light, of the fascinating effect of *slowly* changing forms, of the importance of shadows, and of the fundamentals of abstract design. (Figure 6-2.)

As can be seen in Color Plate X all colors, including white, are obtained from the combined output of the red, blue, and green bulbs. The subtle chromatic modulations produced by the revolving form provide a unique aesthetic experience: even when the turntable is stopped and the form is motionless, an effect of movement is achieved by varying the intensities of the colored lights by use of the dimmers.

Plate IX. Mobile color effects projected on a wall by a simple cove lighting device described in the text.

Great variety of pattern and color distribution is possible through change of position of the colored bulbs, which can be moved about in all directions. This provides for the exercise of choice as to the most satisfactory arrangement of bulbs and the resultant play of lights and shadows.

The bulbs need not be colored, although color is a basic resource of this instrument. Much of composition and design can be learned from working in "black and white," i.e., using ordinary light bulbs. Indeed, just as drawing is traditionally a prerequisite to the study of painting, clear colorless bulbs may be the best to begin with.

The planning and constructing of the forms, with prevision of the kinetics results, is a new and refreshing approach to the study of abstract design.

Although the Sculptachrome served its purpose well, the range of possibilities with a single shadow-casting form was limited, and a great improvement came with the adoption of the compound turntable (Figure 6-3). This turntable consists of an inner disc and an outer rim independently driven by two variable speed reversible motors, which makes possible the use of two forms, one on the disc and one on the rim, each moving at its own slow pace, either clockwise, or counter-clockwise, or not at all—with both forms designed to combine in harmonious relationships (Color Plate X). Figure 6-3 shows the inner disc and outer rim identified as black and as white with their respective shafts and pulleys, but it does not show the motors or their mountings, that being a matter of simple mechanics.

We have now approached infinity in regard to compositional variety, but there is need for something more: the ability to spotlight a center of interest, to create accents here and there, to introduce variety of texture, etc. That calls for devices which would be in the way of the spectator's view and spoil the effect. A new approach must be found; and that approach is the Chromaton.

6-3. A compound turntable as used in Sculptachrome and Chromaton.

7 The Chromaton

CREATIVE EXPRESSION WITH LIGHT

The Chromaton, like the Sculptachrome, had its beginning in the Colortron: and it, too, is very simple in principle: a box or cabinet containing light bulbs; abstract forms in front of these bulbs; and a translucent screen, on the outer side of which the viewer sees the shadows of the forms as they are projected from within.

The original Chromaton (then called the Symphochrome) was simply a new utilization of the Colortron in which, instead of viewing the forms and their shadows from in *front* of the cabinet as in Figure 5-2 of Chapter 5, one viewed the *shadows only* from the *rear* as they were projected through a translucent screen, as in *B* of Color Plate VIII. The point of view was that shown in Figure 5-3 of Chapter 5. Here, without the translucent screen, we see the light bulbs and the star itself. If the screen were placed in the opening, we would see only the *shadow* of the star.

The original purpose of the Chromaton was to provide my design students with a new medium of expression: a means of exploiting the aesthe-

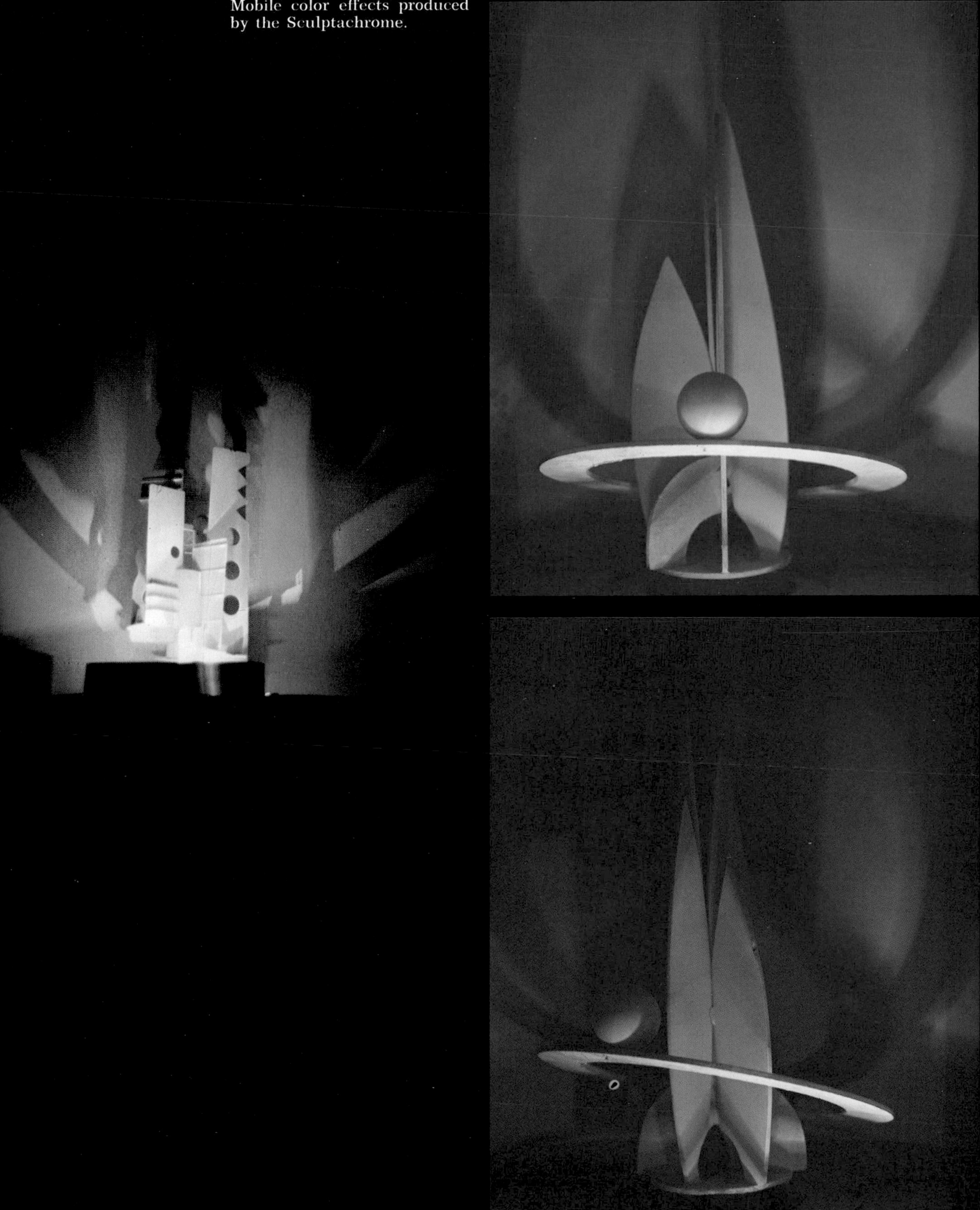

Mobile color effects produced by the Sculptachrome.

tic possibilities of colored light discovered through their experiments with the Colortron, and a means whereby they could express themselves in abstract compositions without advanced skills in drawing and painting. Many of them had ideas far beyond their ability to put on paper or canvas, and they found new encouragement, self-confidence, and enthusiasm in this new key to creativity. Designs created on the Chromaton were *alive* with the intrinsic vitality of light.

The creative activity provided by the Chromaton was in two stages: first, the planning and construction of the abstract forms; second, the use of these forms in the instrument. The first required a new kind of imagination: the ability to visualize what the forms would "do" when their shadows were projected by a single light source onto the screen of the instrument, what they would do when turned in different directions by the turntable, the same compound turntable used in the Sculptachrome (Figure 6-3), and how the shadows of the two forms would interact in various relationships. This could not all be foreseen, of course. There were many happy surprises and many disappointments with the first results, much trial and error and editing of the forms before they were satisfactory.

It was soon realized that just any-old-thing would not do (though even that could be interesting when projected in shadows of several colors, and sometimes gave a clue to a successful design). There must be organization, and the forms must be basic and pure. It was my premise that such forms would always produce pleasing combinations; the results of the experiments seemed to bear this out. Some of these forms are shown in Figures 7-1, -2, -3.

So far we have been concerned with single shadows of the forms cast by a single light source (white or colored) directly behind the form. (Any of the light sources can be any color, depending on what filter is used.)

A great new challenge to talent and taste comes with the use of more than one colored light source with resultant multiple overlapping shadows. Sitting at the remote control panel the student can now bring forth an endless variety of color compositions: he has at his command a "palette" never approached by paint in range or purity of color. By turning various dials he can instantly achieve undreamed-of harmonies, and here he can exercise (and develop) his critical faculty. By recording the readings of the dials he can, at any future time, repeat the effects he has achieved—or he can instantly wipe them out and start over.

The use of music as a stimulus to the creation of mobile color sequences was another intellectual and aesthetic adventure. Music was not only a stimulus to fantasy and invention, but it also provided a framework on which to build. While there was no conscious attempt to translate the pitch and key of the music into parallel expressions in color, a natural response to the overall *mood* of the music induced an identifiable relationship in the resultant color compositions. The music used was always quiet, calm, and slow in tempo.

7-1. Construction form for Chromaton.

7-2. Construction form for Chromaton.

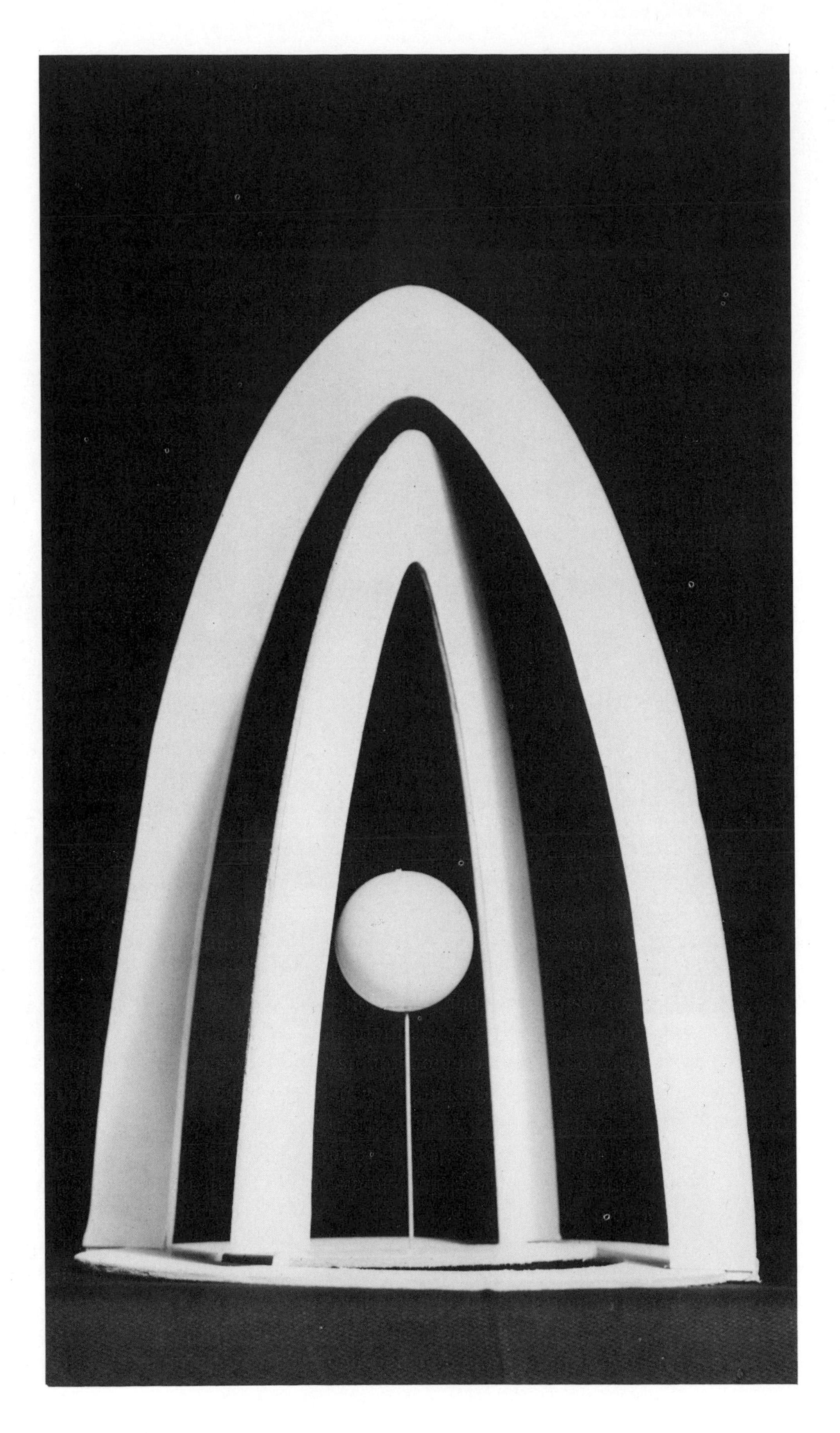

7-3. Construction form for Chromaton.

The results of this new approach to design carried over into the students' work in textile design, package design, graphic art, and other courses. There was a new feeling for color, a new display of imagination, and a new grasp of the essentials of non-objective art.

At this point the Chromaton had outgrown its quarters in the cabinet of the Colortron and had acquired a larger cabinet of its own, and a more sophisticated remote control panel large enough to be called a "console."

Also at this point, the Chromaton had achieved the status of a recital instrument, and I built a portable, collapsible version with which I gave numerous public demonstrations of color and music.

Furthermore, interest in the instrument had spread beyond the academic walls, and there were frequent callers to my "color-machine room" (which the University had generously provided) to see the Chromaton and to "play around" with it themselves. The callers were not all from the world of art: a banker and a business leader found relaxation in working the machine after a grueling day in the marketplace; an electrical engineer and a geologist meticulously prepared forms with which to experiment.

This availability of the Chromaton for use by individuals is, I believe, a new and valuable feature that distinguishes it from instruments that have been designed solely for public demonstration.

The Chromaton in its latest and most sophisticated form is shown in Figure 7-4. In the foreground is the control console, connected by a cable to the projection cabinet, which rests on a pedestal that holds supplies, special effect devices, and forms not in use.

The aluminum cabinet, with the translucent screen at the front, contains the turntables, the upper and lower banks of colored lights, special projectors, spotlights, and the forms whose shadows appear on the screen. There are also outlets into which special-effect devices can be plugged, and for which there are corresponding reserve controls on the console.

The screen, of flashed opal glass, is removable and can be moved forward or back to adjust the size of the projected shadow. Doors at the back give access to the interior of the cabinet.

Three materials have been tried out for use as a screen: tracing vellum, a rear-projection fabric from the Raven Screen Corporation of New York, and flashed opal glass. All gave good results, but for reasons of economy and availability I recommend tracing vellum for the first venture in building a Chromaton. The Raven Screen product is excellent, and the daylight type (black) does not require a totally darkened room (although that is always to be preferred). Flashed opal glass, however, has the advantage of eliminating "hot spots" and of presenting the same image regardless of the observer's angle of view. It also gives an appropriate diffused image and for these reasons is used on the Chromaton.

The console looks forbidding, and it is indeed more complex than is usually necessary. It was deliberately overdesigned, as was the Chromaton

itself, to provide for all possible future needs; many of the controls have been little used.

For a reasonably adequate console the essential electrical equipment comprises:

fourteen solid-state dimmers (one for each of the six colored light sources; one for a master to control the input voltage to all six; two for spotlights; three for special projectors; and two for special effects).

fourteen cartridge fuses and clips (one 5 amp for the master dimmer and thirteen 1 amp for the other dimmers).

thirteen SPST toggle switches (one for the main switch and one for each of the twelve dimmers).

one pilot light and socket.

wire for the cable connecting the console to the cabinet and for wiring within the two.

two 200-watt Variac or other variable autotransformers to control turntable motors.

Those who have built the Colortron should have no difficulty in wiring this console and the cabinet, and in arranging the controls on a panel similar to but larger than that made for the Colortron.

Figure 7-5 is a front view of the cabinet with the screen removed and the rear doors opened. *A* is the pair of forms on the compound turntable *C*. *B* and *B* indicate the two banks of colored light sources—three lights above and three below. The individual lights are adjustable vertically and laterally, and any color filter can be placed in the front of the housing that fits over each individual bulb, thus making it possible to have several arrangements. I usually have red, blue, red, above, and green, blue, green, below. *D* is one of two spotlights, the other being partially concealed behind the form. *E* is a "cloud machine"—a glass cylinder painted with a pattern of stylized clouds revolving very slowly around a small light bulb, when its motor is running. (Sometimes the cylinder is kept stationary and the projected "clouds" are used only to give variety to the background of the image on the screen.)

F and *F* indicate Linnebach (direct beam) projectors, four in all, which are occasionally used to superimpose naturalistic subjects onto the abstract compositions on the screen for surrealistic effects (see Color Plate XI). The projectors are simply tin boxes with small light bulbs inside, and with channels above and below the open front into which any desired photographic or handmade slide can be inserted. Each projector is controlled by a dimmer, and interesting changing montage effects are possible. More will be said about Linnebach projectors later.

7-4. Studio view of Chromaton, with control panel.

7-5. View looking into Chromaton, with front panel removed.

The cove in the center of the top of the cabinet contains several colored bulbs that are occasionally used for special effects.

The components needed for a somewhat simplified cabinet (only two Linnebach projectors) are:

six GE 25T8DC bulbs and sockets.

six handmade filter housings for the above bulbs, with openings and channels in front to receive filters.

three GE PR12 6 volt flashlight bulbs.

three 6-volt bell ringer transformers.

one 1-rpm clock motor (Edmund Scientific Company).

one handmade compound turntable.

two 24-volt DC back-geared type A-9A-621 motors (The Globe Industries, Inc. Dayton, Ohio).

two 500-mA diode rectifiers (any electronic supply company).

two filament transformers, 110-7.8 volt, 40 VA (any electronic supply company).

one cooling fan for colored light housings (Edmund Scientific Company).

colored glass or cellulose acetate filters cut to fit filter housings (Edmund Scientific Company).

one removable flashed opal screen.

A

B

7-6. Colored light units for Chromaton.

C

D

Figure 7-6 shows how the colored light units are made and installed. *A* shows the GE 25T8DC bulb in its socket, which is attached by a bracket to two #372-C "button" magnets (General Hardware Manufacturing Company, New York City, distributors).

B shows the bulb enclosed in a casing of 1¼″ square metal tubing 3¼″ high. In front, the casing has an opening 1⅛″ high and 1⅜″ from the bottom. This opening has a ⅛″ channel at top and bottom for inserting the filter. A glass filter is shown partly inserted.

There are six of these units and six vertical sheet steel plates to which they can be held in place by the magnets. Three of these plates are 6″ high and 3¼″ wide, and three are 3″ high and 2½″ wide. All are provided with bent right-angled bases by which they are fastened to the base of the cabinet behind the turntable.

D shows a taller plate installed 2½″ behind the shorter plate, which should be about 2″ behind the turntable. (This latter distance is variable, since the turntable can be moved forward or back to change the size of the image on the screen.) *D* also shows a light unit in place on the front vertical plate and another on the rear plate, which is directly behind that in front (*C*). There are three sets of these plates, one tall and one short in each set. The taller rear plates are installed 1″ apart.

This system of magnets and steel plates makes it possible to position the light units at any point on the surface to obtain any desired configuration of colored light sources. See Figure 7-7. This shows the bank of colored light units, with filters in place, behind the compound turntable. The units can be raised or lowered and moved to right or left to produce any desired arrangement. Here the base of the bank of light units and that of the turntable are each fitted with wheels which run on tracks, and are individually movable forward or back by remote controls on the console. It is thus possible to regulate the sharpness and size of the image that the forms placed on the turntable will cast upon the screen.

The dimensions given have been carefully worked out and should be strictly followed.

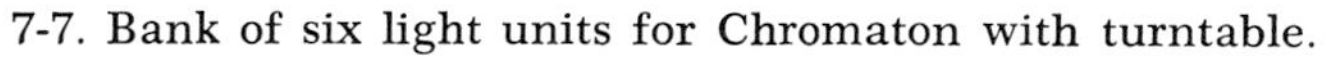

7-7. Bank of six light units for Chromaton with turntable.

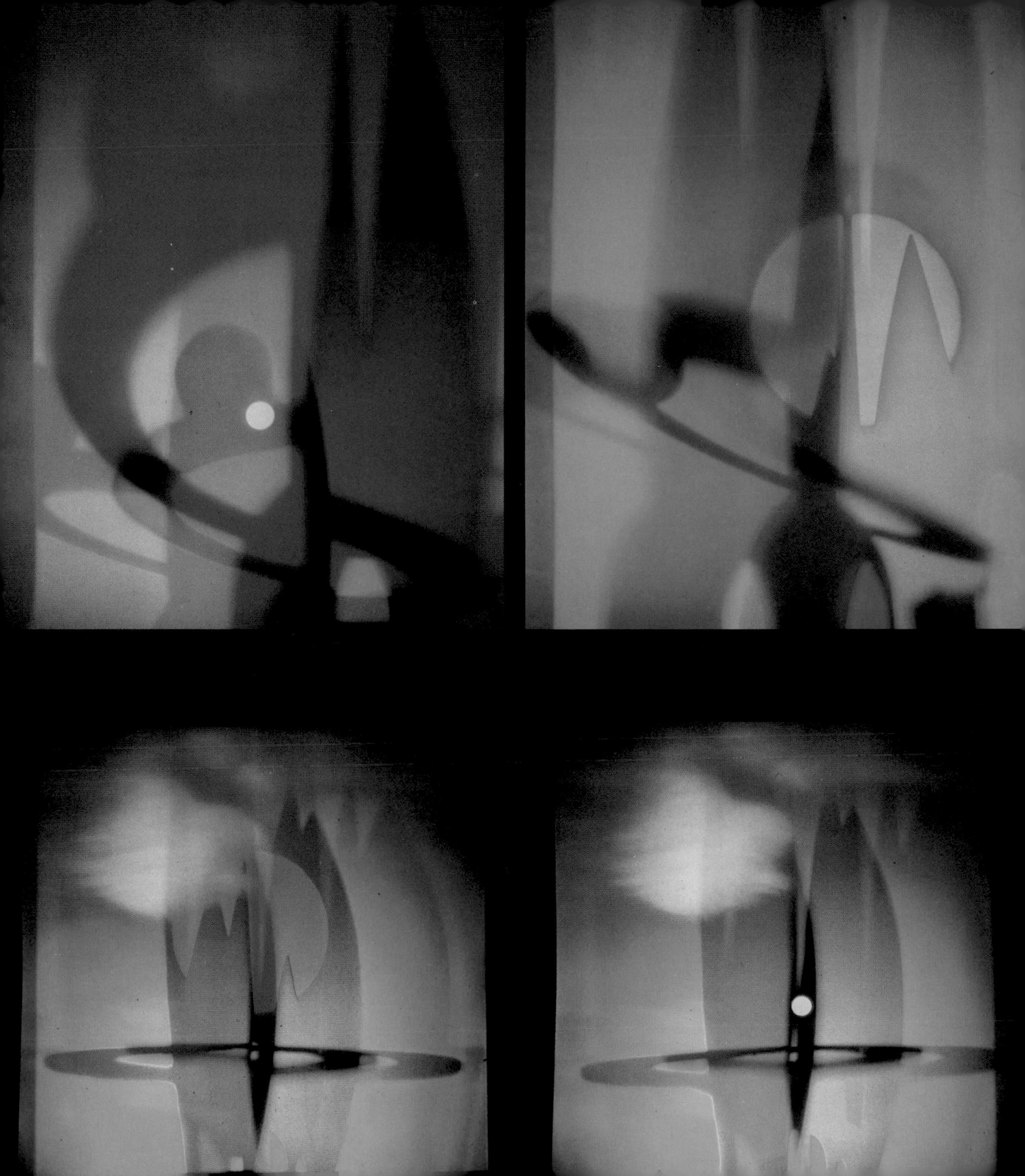

Plate XI. Mobile color effects produced by the Chromaton, an instrument which provides a new approach to creative design.

Figure 7-8 is a wiring diagram for the Chromaton console and cabinet.

The cabinet shown in Figure 7-5 is 28″ wide, 26″ high, and 19″ deep. The screen is 30″ x 30″ without the narrow frame. The details of construction and choice of materials is left to the builders—presumably students.

The rate of rotation of the inner disc and outer rim of the turntable should vary from barely perceptible to not over 5 rpm, and the gear or pulley ratio between them and the motors should be calculated accordingly.

If most or all of the work is done without outside professional help, the cabinet and console should be built for less than five hundred dollars.

For examples of the effects produced by the Chromaton see Color Plates XI, XII, and Figure 7-9 A, B, C, D for similar effects in black and white.

7-8. Chromaton wiring diagram.

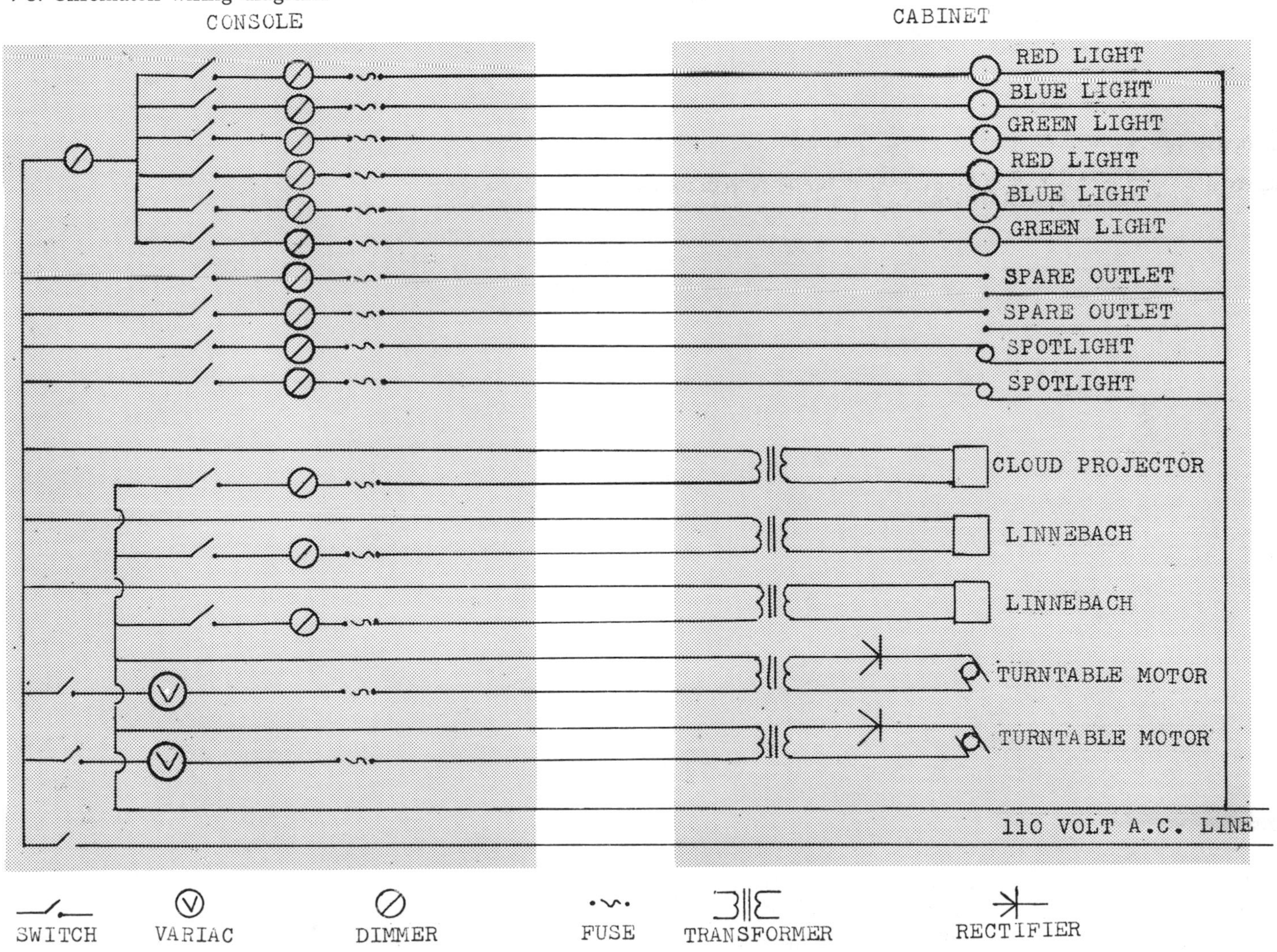

A

B

7-9A, B, C, D. Here are a few of the infinite variety of effects that can be produced on the Chromaton from just one coordinated pair of forms.

C

D

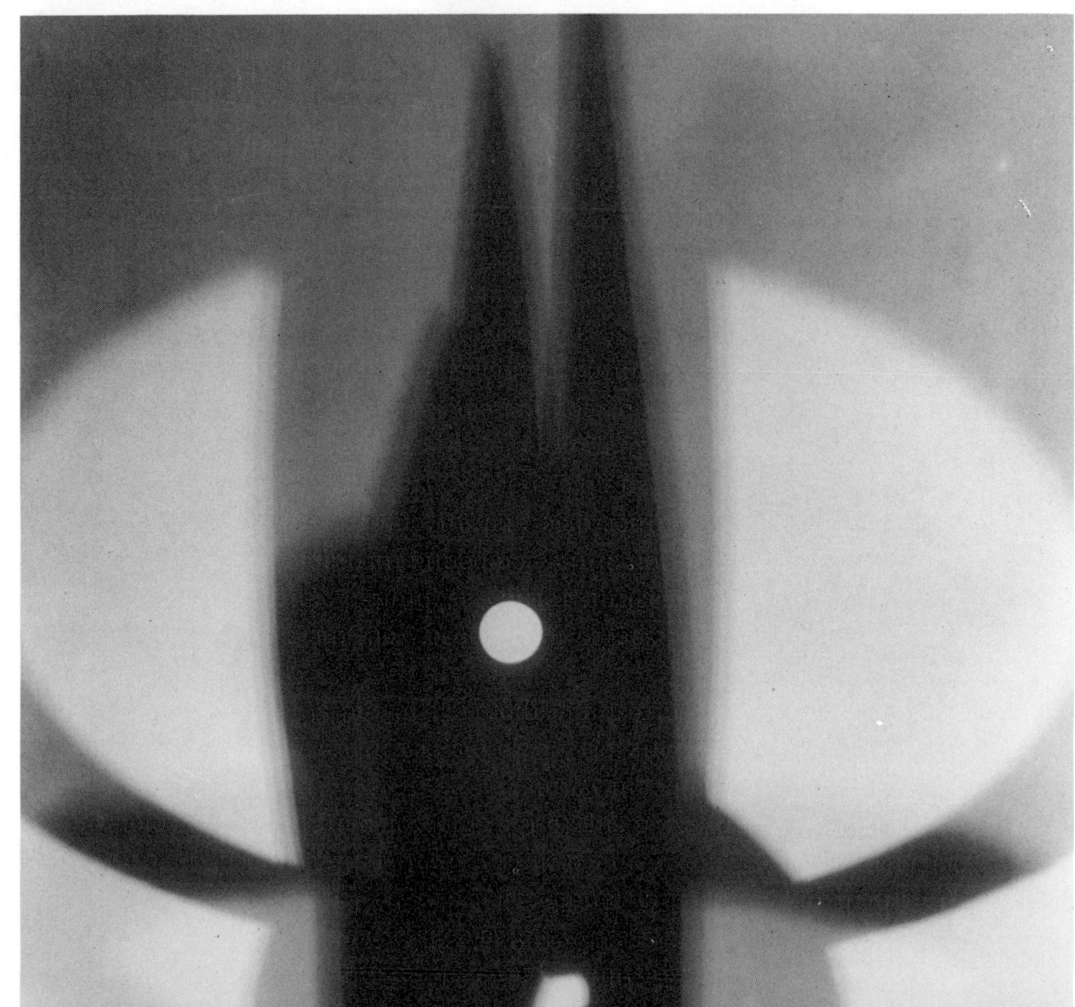

THE LINNEBACH PROJECTOR

Plate XII. More creations from the Chromaton.

The Linnebach projector, so named after its inventor Adolf Linnebach, is used on the stage to project large images onto a white or gray backdrop from a very short distance. It need not be directly in front of the backdrop, but can be off to the side and still project an undistorted image as long as the screen of the projector is parallel with the backdrop. It consists of a large box (two or three feet wide for stage use), a very bright concentrated filament bulb within the box, and a slide of glass or clear plastic in the front of the box. The design on the slide is painted with translucent colors.

The projector, in miniature, is used in the Chromaton as a special-effect device. I see in it possible application in the home (Color Plate XIII) and have made one in suitable size—12″ x 12″ x 10″—which will cover an 8′ x 12′ wall from a distance of six feet.

Figure 7-10 shows a slide (painted in translucent paint), a stand that holds the bulb and transformer, and the box. In use, the stand is placed inside the box and the slide is inserted in the channels in the front. Through an opening in the rear of the box the bulb can be moved up or down and the stand can be moved about until the image on the wall is projected in the desired size and location.

7-10. The Linnebach Projector, showing black box, light source with transformer, and translucent screen.

A

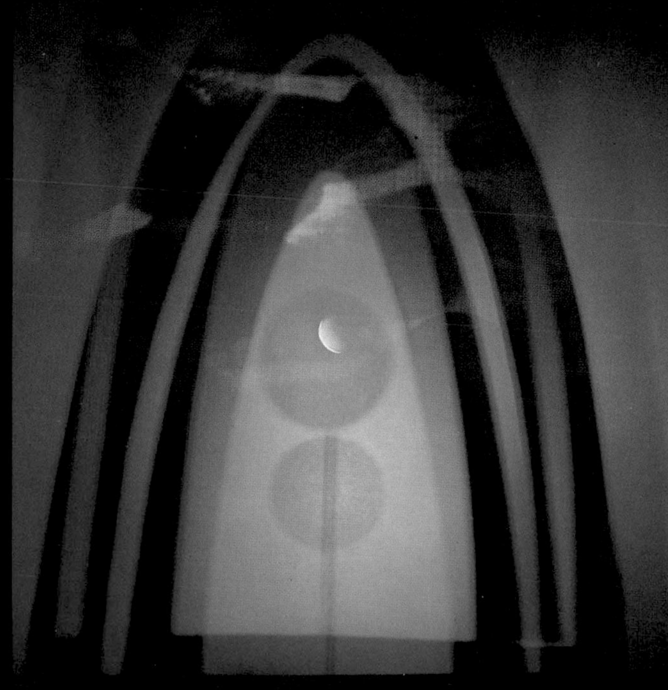

B

C

The transformer is a Sylvania filament transformer 7.8 volts 40VA. The bulb is a 1493 GE 6.5 volts. The excessive voltage from the transformer causes the bulb to burn much brighter than it does with its rated 6.5 volts, which is desirable although the bulb will not last its normal life.

Figure 7-11 shows a small color wheel mounted on a stand in front of a GE 1493 bulb. The color wheel is a 6″ plastic disc on which are cemented three segments of colored filter—blue, pink, and green (any colors can be used)—leaving one segment clear. The disc is mounted on the shaft of a 1-rpm synchronous clock motor (Edmund Scientific Company) and can be rotated to bring any segment in front of the bulb to flood the projected scene with the color of that segment; or it can be let run continuously, bringing interesting color changes to the projected scene and adjacent furnishings. In the effect seen in Color Plate XIII this unit was replaced by that shown in Figure 7-10.

This Linnebach projector could be built for less than twenty-five dollars. Without the color wheel it could be made for less than ten dollars.

7-11. Color wheel, which can be used with Linnebach Projector.

8 The Celeston

AN ADVANCED ART FOR THE FUTURE

In the Celeston, color effects are more or less fluid and abstract and to many observers seem to resemble a kind of "visual music." See Color Plates XIV, XV, and XVI (*frontispiece*).

In fact my approach to the art of mobile color was inspired by years of study of the piano as well as of painting. Primarily interested in the piano, it was natural for me to think of color in terms of music—though I sought no parallel between the hues of the spectrum and the musical scale. Rather, through correspondence of tempo and dynamics (softness and loudness; dimness and light) and the affective influence of colors and appropriate moving forms, perhaps I intuitively transferred the mood of music into the realm of vision. My early procedure was to build a visual accompaniment around the music as it came forth from a record player, playing it over and over, and, by a system of notation, recording the chosen settings of the controls of my instrument as the music progressed. The visual interpretation was thus keyed to the music for repetition at any time, and the Gestalt thus produced was more than the sum of its parts—bringing a new excitement and pleasure to all who saw and heard it; even to those with little feeling for color or music alone.

Decorative effects in the home made possible by the Linnebach projector, an extremely simple device, described in the text.

PLATE XIII

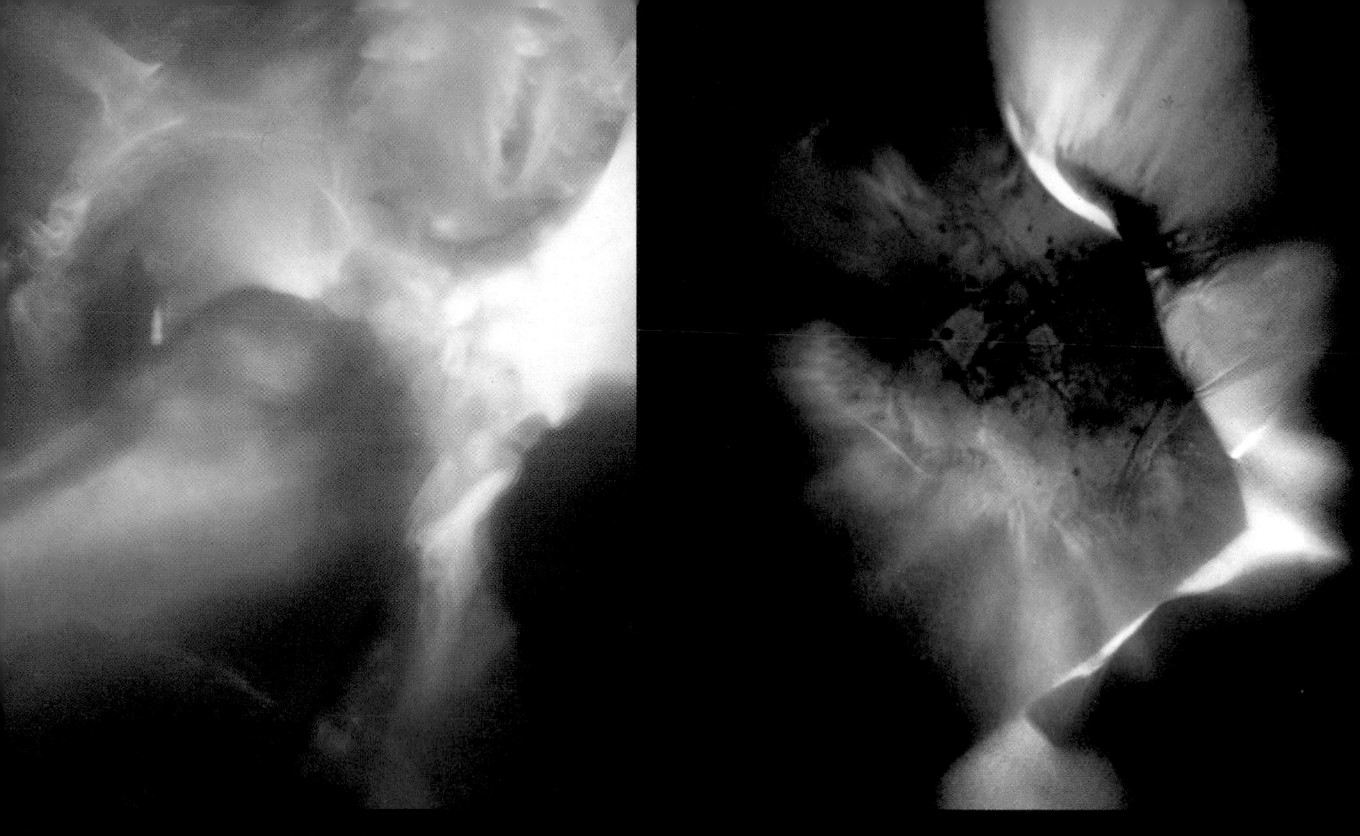

Mobile color effects produced by the Celeston, another creative instrument.

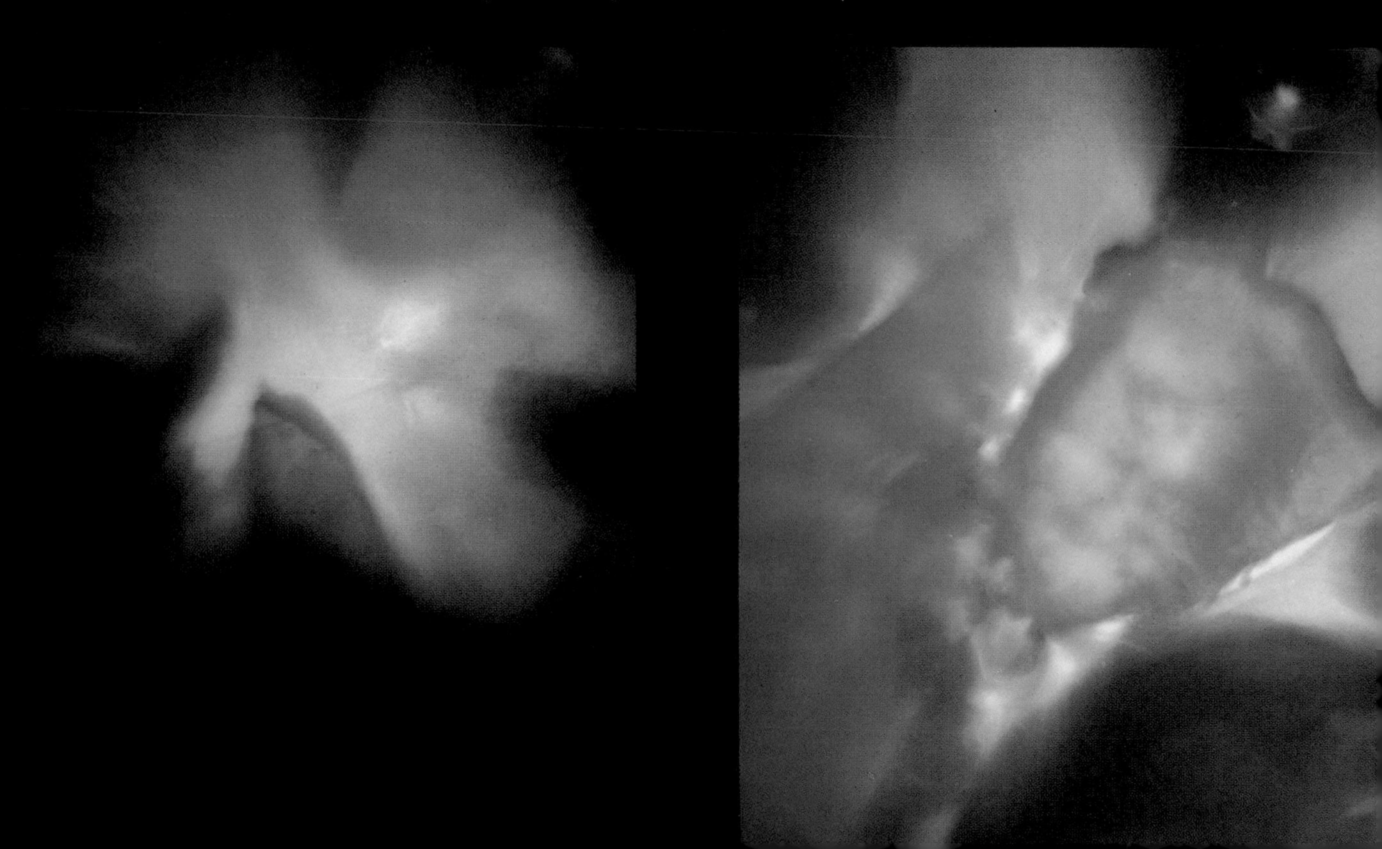

The Celeston, here described and illustrated, was used as a special-effects device in an earlier Chromaton, but it displayed such remarkable capabilities that it has become an important independent instrument. Its basic principle was described by G. A. Shook in 1934, but the machine shown in Figure 8-1 involves several modifications. Figure 8-2 shows the machine with the cover removed, looking downward from the front. *A* is a 300 watt projection bulb. *B* and *B* are two overlapping 6″ glass discs to which are cemented carefully selected and carefully arranged small prisms, colored glass beads, pieces of colored glass and cellulose acetate filter. *C* is a lens, preferably wide-angle (Edmund Scientific Company) with a knob for focusing.

When the bulb is lighted, an image of the bits of colored glass, etc., is projected through the lens onto a wall or screen, and as the discs slowly

8-1. Exterior view of the Celeston.

8-2. Front view of Celeston, with cover removed.

revolve, one in front of the other, very beautiful transitions of form and color appear. The discs are turned by a motor in the base by means of shafts to which they are fixed.

The relationship between the two discs can be altered in two ways: one, by dials *D* and *D*, which move the discs closer together or farther apart; and two, by the large dials *E* and *E* by which the respective discs can be slipped on their shafts. Between these two adjustments infinite combinations are possible, and each adjustment brings new areas of the discs into the path of the light and produces new colors and forms on the screen.

The discs can be rotated in either direction and when a particularly pleasing effect is achieved on the screen the motor can be stopped, and by recording the reading of the dials it is possible to repeat the effect at any time. The rear view (Figure 8-3) shows the motor-driven worm that turns the discs.

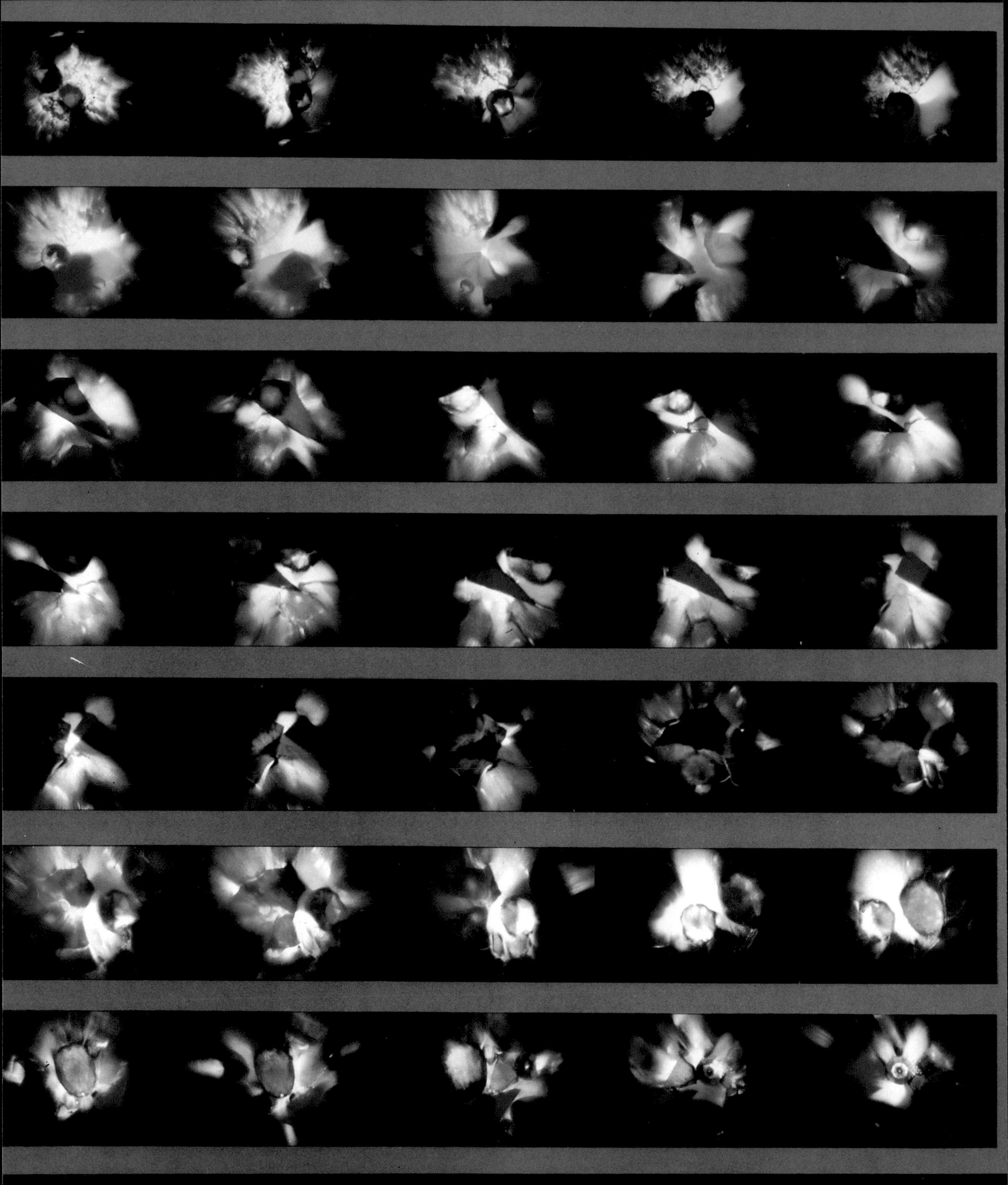

8-3. Rear view of Celeston, showing worm drive.

The creative activity here lies in the making and *editing* of the discs, of which any number of coordinated pairs can be made and fastened easily to the discs.

Plates XIV and XVI (*frontispiece*) show examples of creations by the Celeston. Unfortunately, the element of motion, which is an important asset of the Celeston and the Chromaton, cannot be transferred to the printed page. However, the Celeston sequence shown in Plate XV may be helpful toward a visualization of the effect produced by the machine itself.

The Celeston does not have to be this sophisticated to produce amazing and beautiful results. The adjustments made possible by dials *D* and *D* can be dispensed with, and each disc can be rotated by its own 1-rpm synchronous clock motor (Edmund Scientific Company) through a belt and pulleys. (The ratio should be such that the discs will rotate once in fifteen minutes.)

With these changes it should be possible to build a Celeston for well under fifty dollars.

Plate XV. A time-lapse sequence from the Celeston. This may give some idea of the effectiveness of the slowly changing colors and forms.

9 Color and Psychotherapy

In this writer's opinion, there is going to be vital need for broader uses of lumia in the world of the future. As man spends increasingly more time within controlled environments, as he plans enclosed cities, undersea and underneath structures, as he ventures into space—all of which will take him away from natural surroundings—he must understand and indeed master the elements that have sustained his life. While it is obvious that he must deal with critical problems of air and water pollution, excessive use of chemical fertilizers, insecticides, and disposal of radioactive wastes, *he must also give attention to one of the most vital of all cosmic energies—light.*

The biological influences of light and color have been given extensive study as they affect plants, lowly organisms, insects, birds, and beasts. An outstanding authority, E. F. Ellinger, lists over 4600 references in his book *Medical Radiation Biology*. Universities and laboratories throughout the world are busily engaged in light and color research. The U. S. Atomic Energy Commission has subsidized fifteen different investigations centering around what it terms photobiology. The Argonne National Laboratory in Illinois has a unique biological spectrograph "used in studies of nonionizing radiation (light) upon living organisms."

Light is a sensitive and essential factor in controlling plant growth and gonad (sexual) activity in birds and animals. It affects the glands of the body, the blood, hormones. Yet only lately have differential effects for color in light (i.e., red, yellow, green, blue) been given special attention. Richard J. Wurtman remarks that "one can count on a single hand all of the papers which have attempted to examine which portions of the photic spectrum are biologically active."

However, the research is not quite so restricted. R. van der Veen and G. Meijer found that with some plants red light caused maximum growth and action, followed by the action of blue light, but that the yellow and green were neutral. (Ultraviolet would destroy the plant.) John Ott has noted unusual effects of color in the breeding of mice, mink, chinchillas.

In general he found that bluish light tended to lead to a high ratio of female offspring, pinkish light to a high ratio of male. According to Ott, chinchilla breeders commonly use bluish artificial light in an effort to produce more females.

In one very recent study (1968) sponsored by the Atomic Energy Commission, J. F. Spalding and Associates investigated the influence of color on voluntary activity in albino RF-strain mice. The rodents were placed in cubicles for periods of 18 hours, rested, and then placed for 18 hours in other cubicles until all environments had been tested. The measure of activity was determined by the revolutions of wheels (like those in a squirrel cage). The test resulted in these findings: there was most activity in darkness; next greatest activity was with red. "Activity in yellow light was significantly greater than in daylight, green, blue, and significantly less than in dark and red." Incidentally, blind mice showed little difference in activity, regardless of color, bearing out that the effects discovered were "due to visual receptors."

Mice, of course, are nocturnal animals who busy themselves at night and rest during the day. While their reactions to color can hardly be interpreted in terms of man, one point is significant—mammals definitely do react differently to different colors in an environment, and such automatic reaction has nothing to do with mental response.

Perhaps lumia, the art of light and color, may not be applied in future man-made environments for any direct *biological* values, but surely in the realm of the *psychological* it will be most important. Today man is not only threatened as to the survival of his *body*, but also as to the sanity of his *mind*. People are having mental problems by the thousands. Mental in-

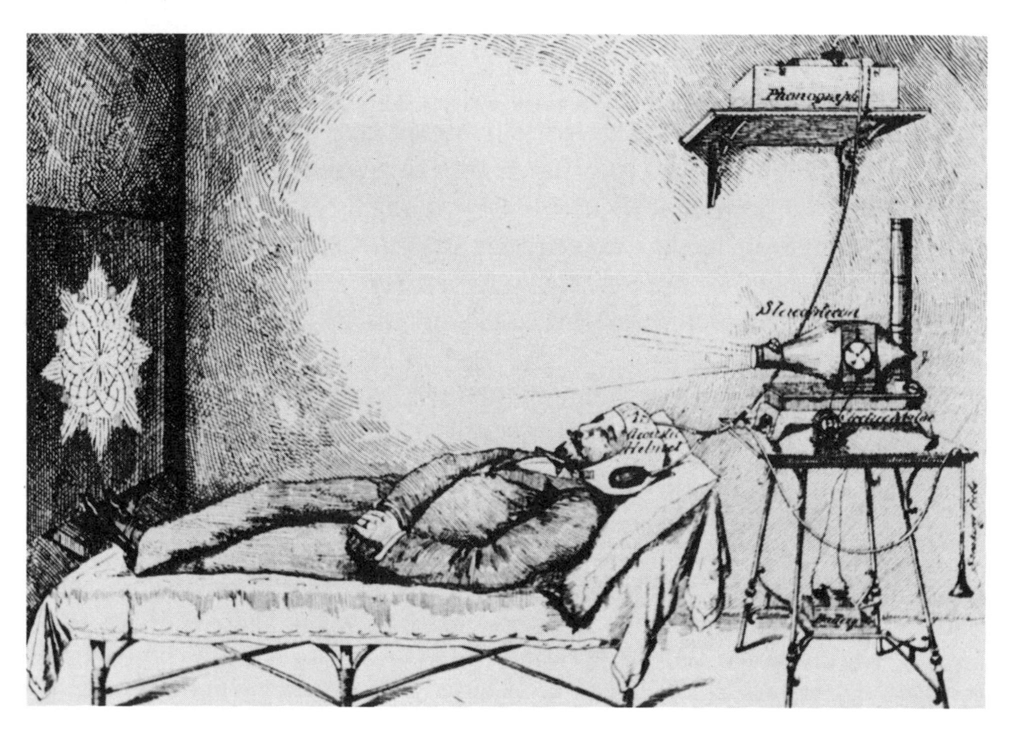

9-1. The Victorian quack used color and music to attempt cures for virtually all human ills, physical and mental.

stitutions are jammed. Mental health centers are opening up everywhere. Carl Jung wisely wrote: "The gigantic catastrophies that threaten us are not elemental happenings of a physical or biological kind, but are psychic events. . . . Instead of being exposed to wild beasts, tumbling rocks, and inundating waters, man is exposed today to the elementary forces of his own psyche." These psychic forces have been encountered by hosts of persons suffering nervous breakdowns, by schizophrenics, and by imbibers of LSD.

Lumia fits in here for what diversion, impact, and equanimity color and light can contribute to badgered mortals.

To discuss two extremely different directions for a possible new world of light and color, in the early forties Cecil Stokes of California created what he called Auroratone films for use in the treatment of mentally disturbed patients. Flowing, abstract forms in full color were thrown upon a screen and accompanied by the singing of Bing Crosby and the music of André Kostelanetz. Sanctuaries of music and color were established in the neuropsychiatric facilities of a number of army hospitals and results documented by Herbert E. Rubin and Elias Katz.(C in Color Plate II.)

Disfigured or crippled patients, not to overlook hosts of other mortals beset by fears and inner terrors, will often resist medical or psychiatric aid. As depression grows worse, serious pathological disturbances may result. What is to be done? In the instance of the colorful films of Mr. Stokes, as reported by Rubin and Katz, "Most patients became more accessible. . . Those whose speech was previously blocked or retarded spoke more freely. . . In this state of accessibility it was possible for the psychiatrist to establish rapport." Here light and color were used for sedative effects.

Today in the modern discothèque, the light festival and light circus, precisely the opposite effects sought by Stokes are being visited upon an excited generation, mostly young. All this may be judged as undisciplined nonsense to the staid of heart, but it is highly meaningful and promising.

Psychedelic lighting and color have not come from the lighting industry but from artists and a few scientists. Anyone who has had his senses accosted will admit that the results were pronounced—either wonderful or awful.

There is just as much color within man as there is in the world beyond. While space-age scientists busy themselves with interplanetary travel, other scientists in psychological realms are equally occupied with inner space. Indeed, man's knowledge of himself, his perception, mind, spirit, has been increased vastly within recent years and vies in magnitude with enlightenment on the physical aspects of the universe.

There is little doubt but that somewhere between the Auroratone films of Cecil Stokes and the discothèque of today lie new functions for light and color in the controlled environments of the future. This future centers around problems being encountered in what is termed sensory deprivation or perceptional isolation.

9-2. Mental illness has become one of the major afflictions of modern civilization. Good color is needed in institutional environments. (United Press International Photo.)

Man has recently been made aware that if the sense of sight (along with other senses) is not stimulated, reactions will take place anyhow. Prisoners in solitary confinement or in prison camps, monks in the seclusion of cells, sailors who venture the oceans alone in small boats, men lost in woods or deserts, are often visited by colorful apparitions for no external cause. Likewise affected are persons confined to iron lungs, or otherwise immobilized by fractures and cardiac disorders. Schizophrenia is also like this. Withdrawn from humanity, hunched in a dark corner, the patient may leave his immediate world for a dream world of his own. As the psychologist R. L. Gregory states, "It seems that in the absence of sensory stimulation the brain can run wild and produce fantasies which may dominate." In the contemporary world, such hallucinations may become an occupational hazard for men who sit at automated machines or who travel into empty space confined to a crowded projectile. Colors and visions from within them may block perception of actual environment.

Equally disturbing is the possibility of a vast neurotic population of human beings cooped up in massive housing projects and plagued by blank walls, a trend which is now taking place and which cannot conceivably be reversed.

In *The Psychology of Perception* M. D. Vernon describes research and clinical studies having to do with sensory deprivation. In one investigation of the effects of a monotonous environment, persons were voluntarily and individually confined to small rooms for periods up to five days. There was little sound; their eyes were covered with translucent goggles to cloud their view, and long cuffs were put over their hands to limit the sense of touch. Not all could endure the isolation for the full five days. All became bored and restless. Most significant, they suffered visual and auditory hallucinations. When they emerged objects appeared blurred, distorted. They experienced dizziness, and when their intelligence was tested it was found to have deteriorated.

In a study of hospitalized infants under seven months of age it was found that a monotonous environment led to confusion and blank stares. As Vernon summarizes, "Thus we must conclude that normal consciousness, perception, and thought can be maintained only in a constantly changing environment. When there is no change, a state of 'sensory deprivation' occurs; the capacity of adults to concentrate deteriorates, attention fluctuates and lapses, and normal perception fades. In infants who have not developed a full understanding of their environment, the whole personality may be affected, and readjustment to a normal environment may be difficult."

There seems to be a connection between the results of isolation and the taking of psychochemicals such as LSD. Woodburn Heron and associates observe, "It appears that exposing the subject to a monotonous sensory environment can cause disorganization of brain function similar to, and in some respects as great as, that produced by drugs or lesions."

Finally, another group of researchers (Herbert Leiderman et al.) have reported on hazards of isolation in medical care. Hospitals in particular need color as well as they need other sensory interests (music, TV, visitors). Leiderman and his group had volunteers willingly confine themselves up to 36 hours in a respirator in which they were able to see only a small area of ceiling. Only five of 17 could endure the confinement for the full 36 hours. "All reported difficulty in concentration, periodic anxiety feelings, and a loss of ability to judge time. Eight of the seventeen reported some distortion of reality, ranging from pseudosomatic delusions to frank visual hallucinations. Four subjects terminated the experiment because of anxiety; two of these in panic tried to release themselves forcibly from the respirator."

What is highly pertinent here is that disturbed or ill people (not to mention sane ones) are often expected to spend long hours and days in confined and drab quarters. Assume that a surgical operation may correct a man's illness or set a man's bones, what then if his confinement leads to other and unexpected maladies? "If normal persons can develop psychotic-like states . . . how much more likely it is that sick patients, perhaps already perilously near the mental breaking point, can be tipped into psychopathological states by the stress of sensory deprivation. Delirium may be imminent for patients weakened by fever, toxicity, metabolic disturbance, organic brain disease, drug action, or severe emotional strain; sensory deprivation may tip the balance. We have accumulated clinical evidence that sensory deprivation may be one element of importance in the etiology of mental disturbance as a complication of various medical and surgical conditions."

Not only hospitals and sanitariums, but convalescent homes, nursing homes, retirement homes need to be planned to combat the frightening dangers of isolation. If old people, for example, can't stand being together with others of their kind—a situation which is usually good for them—and if they prefer solitude, such privacy must of necessity be equipped with colors, sounds, motion, or they will surely encounter neurotic disturbances.

So the art and science of illumination and color should follow up inquiries into things biological with like inquiries into things psychological and psychic.

Taped lumia sequences could be used for diagnostic purposes—like mobile or animated Rorschach cards. Lumia projections on walls, ceilings, panels, designed for excitation or pacification as circumstances suggest, could overcome sensory deprivation and keep a lot of persons sane. With lumia and colored light there could be movement, change, variety, color. There could be programmed light, shifts from soft pink and orange to yellow and cool white of high brilliance. There could be mobile color, kinetic color, the electric circus domesticated for average and everyday habitats—light and color for the body, the eye, the mind, and the soul.

(This chapter has been based largely on an article by Faber Birren, which appeared in the May, 1969, issue of *Illuminating Engineering*.)

10 Color, Music, and Sound Relationships

This book agrees and assumes that there is no scientific relationship between the vibration frequencies of color and the vibration frequencies of sound. In fact, color wavelengths or frequencies are natural phenomena, to be objectively measured. The diatonic scale in music, however, is an arbitrary concept of Western culture, and even though certain frequencies bear relationship, one to the other, any possible alliance with color would be presumptuous.

Yet this is not so with reference to emotional and psychological associations, to what persons may "feel" about color and sound.

In what is known as synesthesia, or color-hearing (there are other forms of synesthesia as well), many odd opinions have been expressed in the past. In a program for the Boston Symphony Orchestra, some years ago, Philip Hale commented on a few of the color associations of musicians. He told that Raff held the tone of the flute to be intensely sky-blue. The oboe was clear yellow, the trumpet scarlet, the flagelot deep gray. The trombone was purplish to brownish, the horn greenish to brownish, the bassoon, a grayish black. He remarked that A major was green to one musician and that another felt the hue of the flute to be red rather than blue, as with Raff. In 1890 a woman was found to whom the music of Mozart was blue, that of Chopin yellow, and that of Wagner a luminous atmosphere with changing colors. To another subject *Aïda* and *Tannhauser* were blue, while the *Flying Dutchman* was a misty green.

This alliance of color with various instruments has been encountered in many persons. Christopher Ward wrote, "From the faintest murmur of pearl-gray, through the fluttering of blue, the oboe note of violet, the cool, clear wood-wind of green, the mellow piping of yellow, the bass of brown, the bugle-call of scarlet, the sounding brass of orange, the colors are music." Albert Lavignac in an article for *Music and Musicians* in 1905 presented the list given at the end of this chapter.

Often colors are associated with the sound qualities of things other than musical instruments. E. G. Lind writes: "We have still another instance of colored sound recorded by a physician . . . who declares he can distinctly see the colors of sounds emitted by the human voice. He says they are red and blue, black, tan, slate, and all other colors, and that the color of some handsome women's voices is like buttermilk; not a very poetic tint but near enough to suggest milk and honey."

Further, Lind held associations of his own. He thought "Yankee Doodle" and "Dixie" to be decidedly American—red, white, and blue throughout. "God Save the King" was typically English in "look" as well as sound. In "Auld Lang Syne" he detected a strong resemblance to plaid! However, of "St. Patrick's Day" and "The Wearing of the Green" he said, "Our prejudices in favor of green are ridden over rough-shod by the prominent, and so more powerful, orange color, a most unexpected and unlooked-for result."

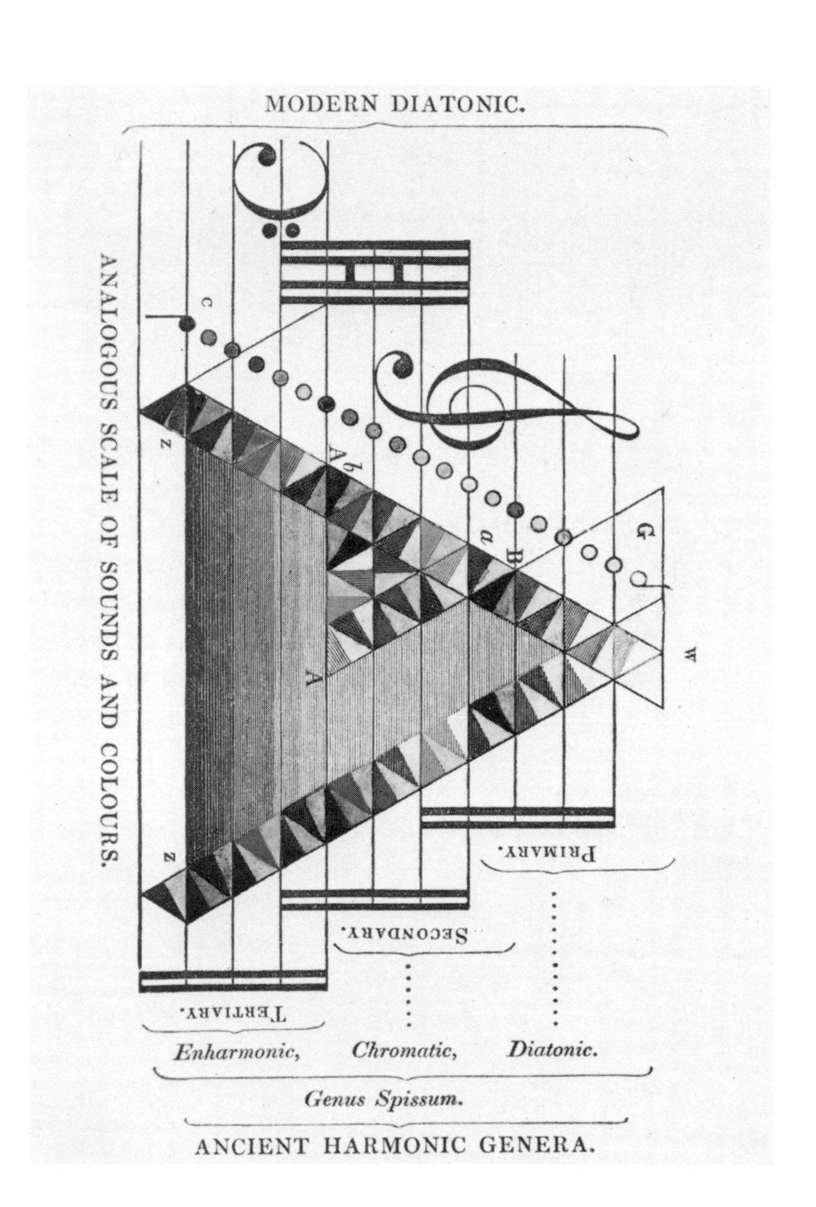

10-1. Colors and the notes of music as conceived by George Field in 1845.

Among musicians, Liszt is credited with a number of pet phrases: "More pink here, if you please." "That is too black." "I want it all azure." Beethoven is said to have called B minor the black key. Schubert likened E minor "unto a maiden robed in white with a rose-red bow on her breast." One Russian composer said, "Rimsky-Korsakoff and many of us in Russia have felt the connection between colors and sonorities. Surely for everybody sunlight is C major and cold colors are minors. And F-sharp is decidedly strawberry red! Of his subtle compositions Debussy wrote: "I realize that music is very delicate, and it takes, therefore, the soul at its softest fluttering to catch these violet rays of emotion."

Klein believes that the idea of true orchestral "tone-color" was perhaps not well understood until the time of Wagner. The great German frequently had color in mind and drew musical pictures with it. He wrote to Frau Willie, "I am differently organized from other men. I must have beauty, color, light."

Wagner in speaking of Auber said: "Even as the subject lacked nothing of either the utmost terror or the utmost tenderness, so Auber made his music reproduce each contrast, every blend in contours and colors of so drastic, so vivid a distinctness, as we cannot remember to have ever seen before. We might almost fancy we had actual music-paintings before us, and the idea of the picturesque in music might easily have found substantiation here."

About one of his own compositions he said: "Here the trumpets and kettle-drums, which for two bars long have filled the whole with splendor, pause suddenly for close upon two bars, then re-enter for one bar, and cease again for another. Owing to the character of these instruments, the hearer's attention is inevitably diverted to this color-incident, inexplicable on purely musical grounds, and therewith is distracted from thc main affair, the melodic progress of the bases."

In Part II of his *Opera and Drama* he wrote: "To music alone was it reserved to represent this stuff to the senses also, namely, by an outwardly perceptible motion; albeit merely in this wise, that she resolved it altogether into moments of feeling, into blends of color without drawing, expiring in the tinted waves of harmony in like fashion as the dying sun dissolves from out the actuality of life."

Much of Wagner's music was color translated into sound. His "fire" music, for example, was meant to interpret the crackling and flickering of flames. In one of his overtures he used different hues to mark his score —red for strings, green for wood-winds, black for brass.

Yet Wagner apparently refused to recognize synesthesia as a part of his own make-up. He wrote: "I have met intelligent people with no sense at all of music, and for whom tone-forms had no expression, who tried to interpret them by analogy with color-impressions; but never have I met a musical person to whom sounds conveyed colors, excepting by a figure of speech."

The composer Alexander Scriabin, however, devoutly allied hue with sound and actually prepared scores having a color accompaniment. His sense of color-hearing was unquestionably vital and real to him. Writing of this strange faculty Dr. C. S. Myers, a psychologist who talked with Scriabin, said, "Scriabin's attention was first seriously drawn to his colored hearing owing to an experience at a concert in Paris, where sitting next to his fellow countryman and composer Rimsky-Korsakoff, he remarked that the piece to which they were listening (in D major) seemed to him yellow; whereupon his neighbor replied that to him, too, the color seemed golden. Scriabin has since compared with his compatriot and with other musicians the color effects of other keys, especially B, C major, and F-sharp major, and believes a general agreement to exist in this respect. He admits, however, that whereas to him the key of F-sharp major appears violet, to Rimsky-Korsakoff it appears green; but this derivation he attributed to an accidental association with the color of leaves and grass arising from the frequent use of this key for pastoral music. He also allows that there is some disagreement as to the color effect of the key of G major. Nevertheless, as is so universally the case with the subjects of synesthesia, he believes that the particular colors which he obtains must be shared by all endowed with colored hearing."

10-2. Score of Scriabin's "Prometheus." Part for a color-organ is at top and called Luce.

A natural mystic, Scriabin devoted much time to color-music analogies. In writing "Prometheus," he adopted a unique color scale which, according to Klein, "has no order from the spectroscopic point of view if written in the order of the chromatic scale, but which assumes an approximately spectral order if we commence with the note C and proceed in the 'circle of fifths.' "

C — Red
G — Rosy orange
D — Yellow
A — Green
E } Pearly blue
B } The shimmer of moonshine
F♯ — Bright blue
D♭ — Violet
A♭ — Purple
E♭ } Steely with the
B♭ } glint of metal
F — Dark red

"Prometheus," the Poem of Fire, was originally played at Moscow and Petrograd. It was perhaps the first musical score ever written to include a part for the color organ. This part, called "Luce," headed the page and was written to blend the hues of his color scale with the concords of the music itself. Later the music was presented in America. The performance took place in darkness. As the orchestra played, colored lights were thrown upon a screen. But the critics were not pleased. One wrote: "It is not likely that Scriabin's experiment will be repeated by other composers."

Yet without the actual use of color or colored light, the English composer Arthur Bliss wrote "A Colour Symphony" having four movements. Here Bliss sought to convey the musical and emotional impressions of four colors.

"I. Purple: The coulours of Amethysts, Pageantry, Royalty and Death.

"II. Red: The Colour of Rubies, Wine, Revelry, Furnaces, Courage and Magic.

"III. Blue: The Colour of Sapphires, Deep Water, Skies, Loyalty and Melancholy.

"IV. Green: The Colour of Emeralds, Hope, Youth, Joy, Spring and Victory."

The symphony was written in 1922, first presented in Gloucester Cathedral, and revised by the composer in 1932.

In the field of painting, Robert Strübin, a Swiss pianist turned artist, devoted the last years of his life to the painting of music in terms of color. To him each sound had an exact color and an exact shape. His compositions looked like music scores, having a beginning and an end and resembling rather stiff charts consisting of horizontal rows of short colored bars.

Another painter of music was I. J. Belmont of America. He paid tribute to Wagner who had a "mighty dream of a union of the arts." Belmont wrote of having experienced synesthesia at an early age and of seeing color as he heard music. He explained, "I use my pigments musically, employing the colors of the spectrum to correspond with the musical notes. Where there are contrapuntal effects in the music, I reproduce this too on my canvas by playing color against color, which is like playing note against note."

Belmont's paintings have a fanciful and spiritual quality, show blends of color, with indications of figures in the use of Schumann's "Traümeri," stallions for Wagner's "Die Walkure." These are echoes of the luminous effects achieved by Turner.

In the medium of abstract motion pictures, Mary Ellen Bute of New York has delightfully interpreted music with the double impact of color and animated motion, having the eye "see" the music concurrently as it is played. Walt Disney did likewise in his animated movie "Fantasia," using the vibrations of sound as they might be seen in an oscilloscope, but adding color and letting the man-drawn fluctuations cavort in amusing ways.

10-3. Frame from non-objective motion picture to accompany music, "Mood Contrasts," by Mary Ellen Bute.

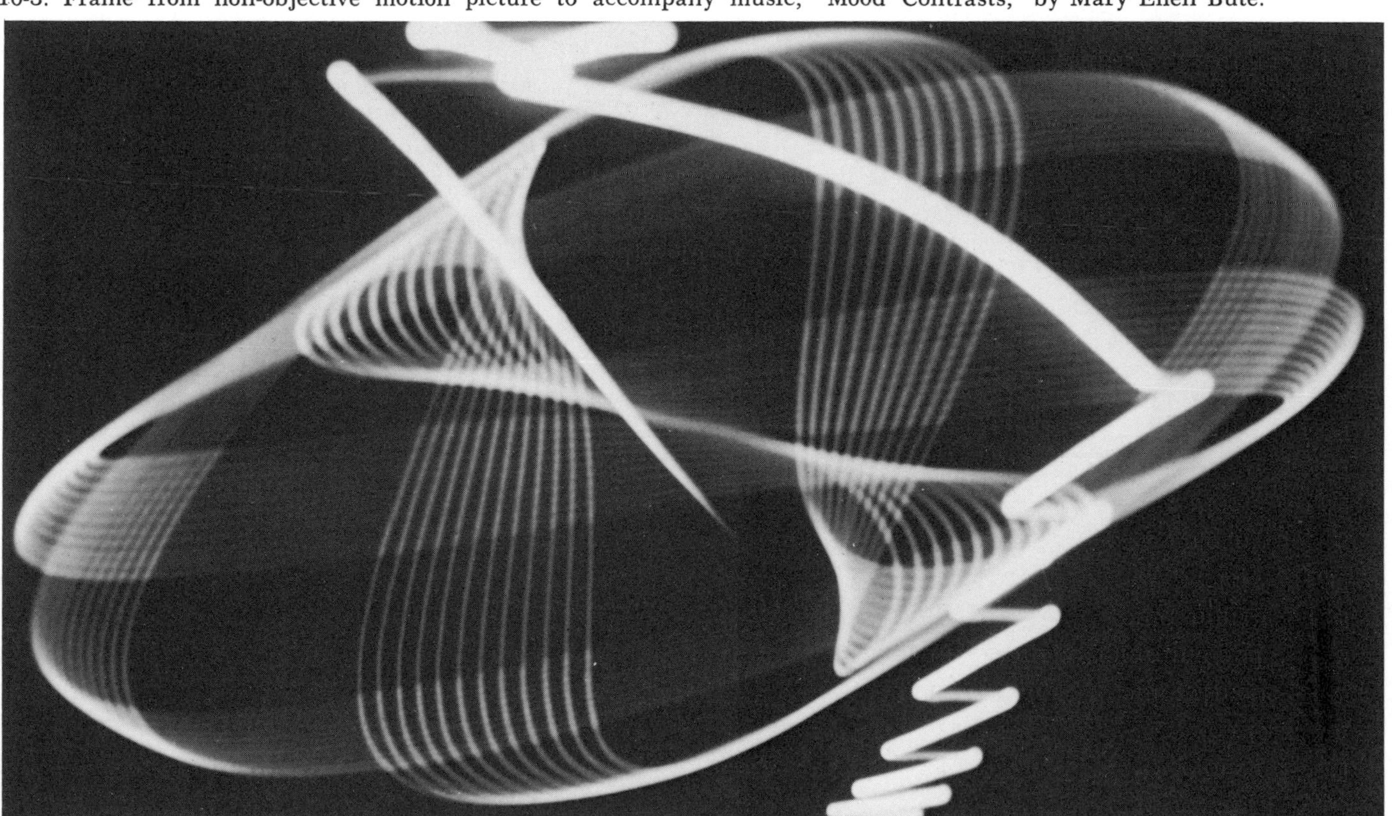

Some painters have knowingly or unknowingly conveyed impressions of sound with color and forms in purely static compositions. Piet Mondrian, the Dutch painter, in his BROADWAY BOOGIE-WOOGIE used red, yellow, blue, and light gray on a white ground in small, geometric staccato squares to suggest the sharp sounds and colors of the "Great White Way."

10-4. BROADWAY BOOGIE-WOOGIE, by Piet Mondrian. Conceived to look like Broadway sounds. (Collection, Museum of Modern Art, New York.)

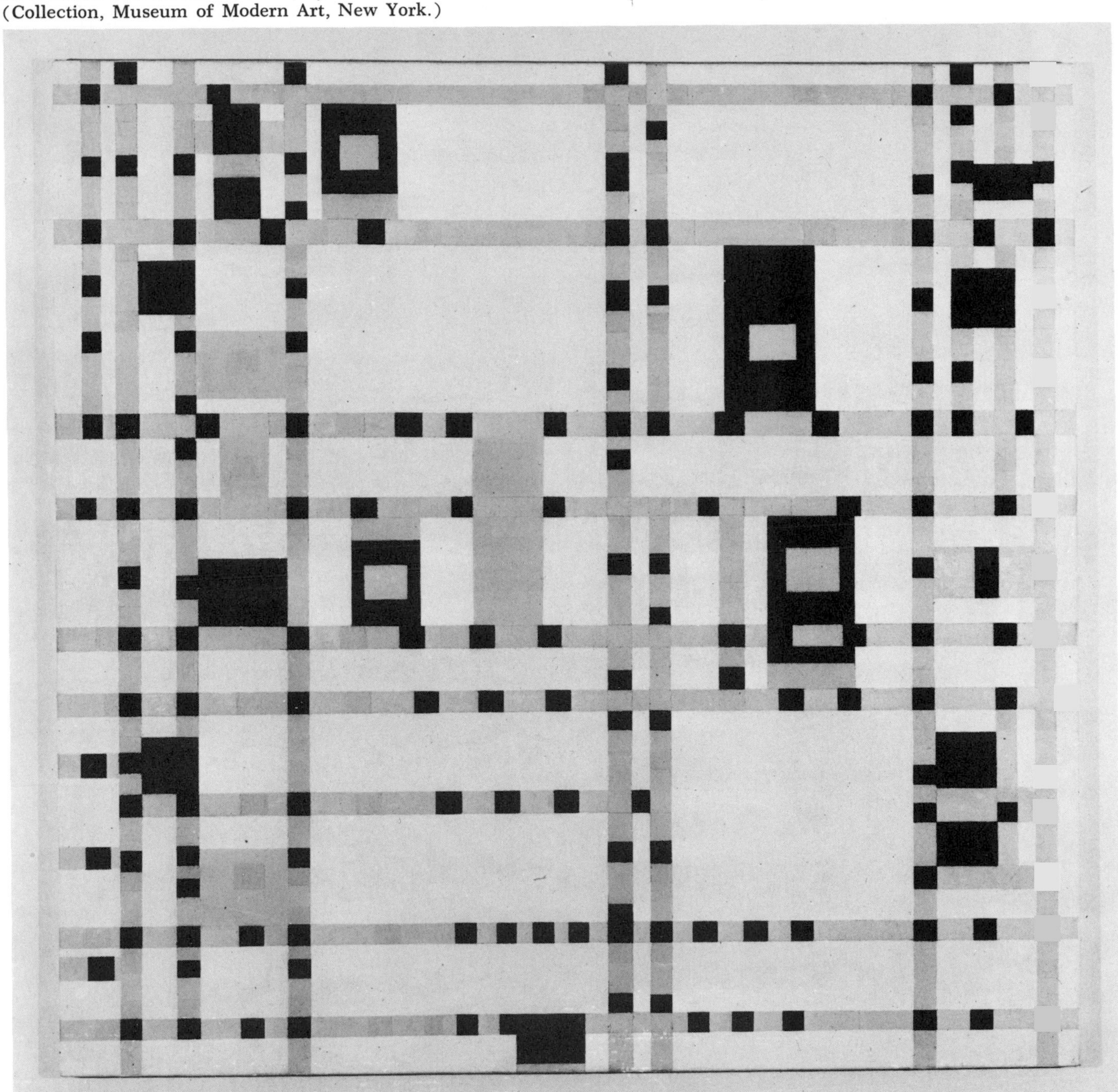

10-5. TWITTERING MACHINE, by Paul Klee. Lavender colors for peeping sounds. (Collection, Museum of Modern Art, New York.)

Similarly, the Swiss painter Paul Klee in his TWITTERING MACHINE used bird-like drawings and soft tones of orchid, pink, and lavender to have his composition "sound" like gentle, chirping peeps.

10-6. ABSTRACTION, 1923, by Wassily Kandinsky. He saw many relationships between color and music. (Collection, Museum of Modern Art, New York.)

A vital love for color and music was held by the eminent abstract painter Wassily Kandinsky, a friend of Paul Klee. In his book, *The Art of Spiritual Harmony,* there are many references. Several are given here.

"The sound of colors is so definite that it would be hard to find anyone who would express bright yellow with bass notes, or dark lake with the treble."

"A parallel between color and music can only be relative. Just as a violin can give warm shades of tone, so yellow has shades, which can be expressed by various instruments."

"To use again the metaphor of the piano, and substituting form for color, the artist is the hand which, by playing this or that key (i.e., form), purposely vibrates the human soul in this or that way."

On the "inner sound" of red: "This inner sound is similar to the sound of a trumpet or an instrument which one can imagine one hears when the word 'trumpet' is pronounced. This sound is not distorted; it is imagined without the variations that occur depending upon whether the trumpet is sounded in the open air, in a closed room, alone or with other instruments, if played by a postilion, a huntsman, a soldier, or a professional."

"Light warm red has a certain similarity to medium yellow, alike in texture and appeal, and gives a feeling of strength, vigor, determination, triumph. In music, it is a sound of trumpets, strong, harsh, and ringing."

"The pure, joyous, consecutive sounds of sleigh bells are called in Russia 'raspberry jingling.' The color of raspberry juice is close to . . . light, cool red."

"In this lies the great difference between a deepened red and a deepened blue, because in red there is always a trace of the material. Corresponding in music are the passionate, middle tones of a cello. A cool, light red contains a very distinct bodily or material element, but it is always pure, like the fresh beauty of a young girl's face. The ringing notes of a violin exactly express this in music."

"The vermilion now rings like a great trumpet or thunders like a drum."

On orange: "Its note is that of a church bell (the Angelus bell), a strong contralto voice, or the largo of an old violin."

"Keen lemon yellow hurts the eye as does a prolonged and shrill bugle note to the ear, and one turns away for relief to blue or green."

"In music, absolute green is represented by the placid, middle notes of a violin."

"In music, a light blue is like a flute, a darker blue a cello; a still darker the marvelous double bass; and the darkest blue of all—an organ."

On violet: "In music it is an English horn, or the deep notes of woodwinds."

White is "like the pauses in music that temporarily break the melody."

Of black: "In music it is represented by one of those profound and final pauses, after which any continuation of the melody sees the dawn of another world."

In more rational terms, color hearing has been more exhaustively treated in a monograph by Theodore F. Karowski and Henry S. Odbert, *Color Music*. According to these writers while synesthesia may have no consistency from one person to another, general reactions to musical phrases are quite uniform. In a study of 148 college students, 39 per cent were able to *see* a color or colors, 53 per cent were able to *associate* a color, and 31 per cent *felt* a color response. At least 60 per cent gave some kind of color response when they heard music. "It seems safe to say that a good majority of the population in one way or another relates colors to music."

Karowski and Odbert found that slow music was generally associated with blue, fast music with red, high notes with light colors, deep notes with dark colors, and that patterns as well as hues were involved. A graphic portrayal of this color and pattern tendency in synesthesia is presented here in diagram form. (Figure 10-7.)

Perhaps further investigation of this sort may lead to new dimensions for an independent art of color-music. In describing forms, an unusual reference is made by Karowski and Odbert to time and space conceptions that seem to have real logic to them. "The horizontal dimension might be related to the development of music in time; the vertical dimension to changes in pitch. A third dimension of depth may eventually be available to denote volume or intensity." Surely most people who have given thought to correlation of music, form, and color will be in sympathetic agreement with this conclusion.

10-7. A diagram illustrating the general nature of uniformities between the various kinds of pattern tendencies in color-hearing. (From *Color-Music*, by Theodore F. Karowski and Henry S. Odbert.)

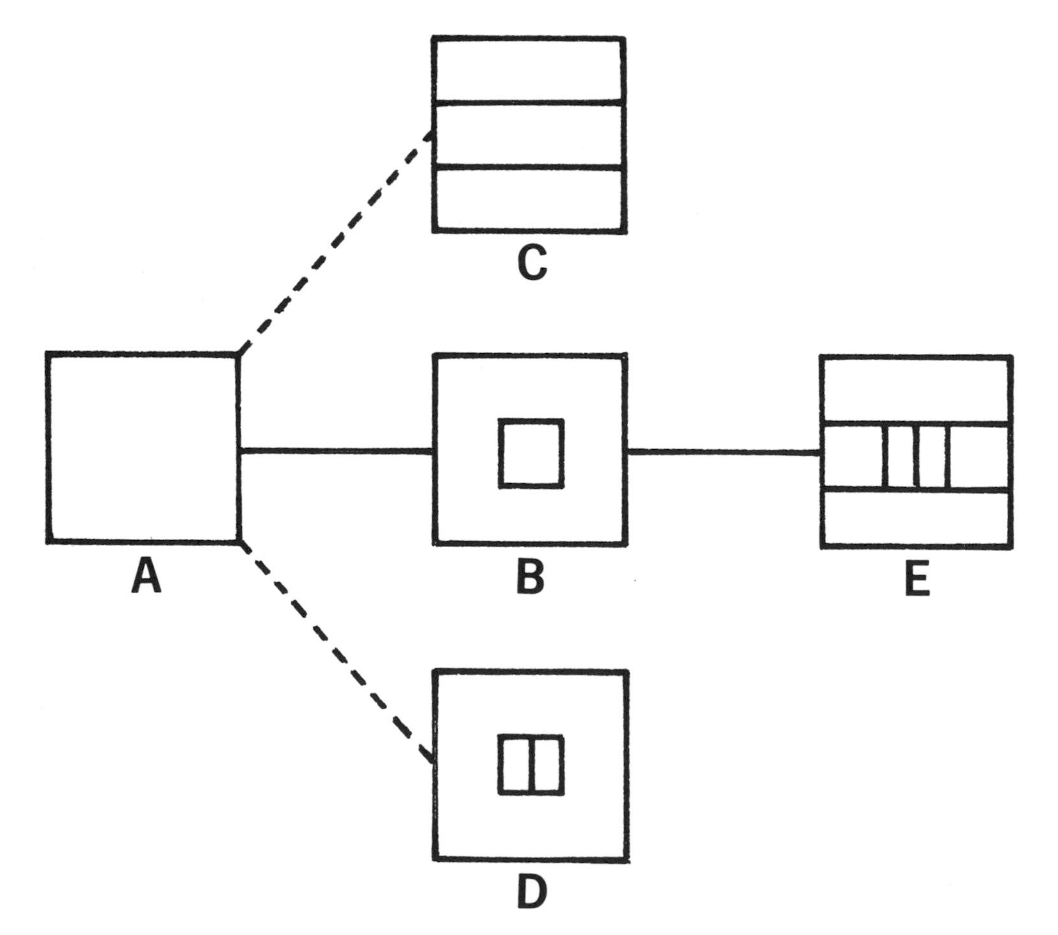

A. Simple color response to musical selection.

B. Simple figure is seen on simple ground.

C. The ground is seen as graded.

D. The figure is seen as graded.

E. Both the ground and figure seem graded—this reaction being found in the more complex cases of synesthesia.

COLORS AND MUSICAL INSTRUMENTS

(After Albert Lavignac)

	Analogous to:
Flute	blue of the sky.
Oboe	a crude green tint.
Clarinet	red-brown, Vandyke red, garnet.
Horn	a brilliant copper yellow.
Cor Anglais	violet.
Trumpets	crimson or orange.
Clarions	crimson or orange.
Trombones	crimson or orange.
Cornet	ordinary red, ox-blood.
Bassoon	grayish dark brown.
Kettle-drums	black.
Side-drum	a grayish neutral.
Triangle	a silvery hue.
Violin (harmonics)	blue.
(IV string)	grave red-brown.
(Pizz)	little specks of black.

COLOR AND ANALOGIES IN MUSIC

	C	**C♯**	**D**	**D♯**	**E**	**F**
Newton and Lind	Red		Orange		Yellow	Green
Castel	Blue	Blue-green	Green	Yellow-green	Yellow	Yellow-orange
Field	Blue		Violet		Red	Orange
Rimington	Deep red	Crimson	Orange-crimson	Orange	Yellow	Yellow-green
Klein	Dark red	Red	Red-orange	Orange	Yellow	Yellow-green

	F♯	**G**	**G♯**	**A**	**A♯**	**B**
Newton and Lind		Blue		Indigo		Violet
Castel	Orange	Red	Crimson	Violet	Pale violet	Indigo
Field		Yellow		Yellow-green		Green
Rimington	Green	Bluish green	Blue-green	Indigo	Deep blue	Violet
Klein	Green	Blue-green	Blue	Blue-violet	Violet	Dark violet

Appendix: General Reference Sources

The following organizations and sources offer products and services related to the art of color and light.

Bausch & Lomb, Dept. 6606, 635 St. Paul St., Rochester, N. Y. 14602. Science Teaching Aids, L-225, describes B & L teaching aids and demonstration kits explaining light phenomena. Material relating to basic and advanced demonstrations of the polarization of light includes polarizers, polarized light materials, and text explaining numerous experiments, many related to color.

Color-Tel Corporation, 13720 Riverside Drive, Sherman Oaks, California 91403. Based on the phenomenon of the Benham disk, this company offers equipment to convert black and white television images to color through the mysteries of subjective (induced) color. An electronic color kit adds a color translator to a black and white TV camera, and a compensating filter to the TV receiver or monitor.

Edmund Scientific Co., 300 Edscorp Building, Barrington, N. J. 08007. A leading supplier of equipment and gadgets associated with light, color, and sound demonstrations, with heavy emphasis on entertainment rather than art. Parts, components—and complete kits—are available for unique visual effects, musicvision, light shows, color wheels, color organs, projectors. Quite valuable are motors, color filters, dimmers, polarizing materials, fluorescent materials, light sources.

General Electric, Lamp Department, Nela Park, Cleveland, Ohio 44112. An important maker of colored light sources. Although incandescent lamps using transparent colored glass are expensive and gradually being discontinued, General Electric has created excellent substitutes, such as Lexon colored sign lamps, colored reflector lamps, tinted coloramic lamps (vitreous coated), and the remarkable Dichro-Color lamps which employ interference filters for the efficient emission of rich, intense colors. Small and large natural bulbs which can be used in combination with color filters are available in generous variety.

Intermedia Systems Corporation, 711 Massachusetts Ave., Cambridge, Mass. 02139. This organization, to a large extent inspired and guided by Gerd Stern (mentioned in Chapter 3 of this book), offers a complete and sophisticated service in the field of color, light, and sound programming. Various audio-visual control panels and consoles are available which will operate (manually or automatically) slide and movie projectors, lights, curtains, display signs, stereo speakers and the like to suit any need, commercial, educational, entertainment or otherwise. In addition, Intermedia personnel will design programs on a custom basis for "encoding magnetic tape, mixing sound tracks, and preparing punched tape programs for clients."

Kodak Motion Adapter. Intended for use with Kodak Ektagraphic Slide projector, but can be used also with any Kodak Carousel projector by attaching a special adapter bracket. A revolving polarizing disk will reveal motion and color change in special 35mm polarizing slides—such as Technamation slides (see reference). Available through Motion Picture and Education Markets Division, Eastman Kodak Co., Rochester, N. Y. 14650.

Lighting Associates, Inc., 351 East 61st Street, New York, N. Y. 10021. Jackie Cassen and Rudi Stern, both mentioned in Chapter III, have done much original work in the field of kinetics and environmental light. They offer a design service through the above company. They will create and install kinetic light effects for building interiors and exteriors. "A room environment created by Cassen/Stern is an artistic expression that brings together objects and people in a harmony of changing, moving color, light and sound."

Polacoat, Inc., 9750 Conklin Road, Cincinnati, Ohio 45242. Efficient light polarizing filters and excellent rear projection screens are offered and are of top professional quality.

Stroblite Co., Inc., 29 W. 15th Street, New York, N. Y. 10011. One of the best sources for ultra-violet products and equipment, fluorescent paints, inks, chalks and crayons, invisible inks, luminous paints, blacklight lamps, strobe lights—all well suited to theatrical uses.

Technamation, Inc., 30 Sagamore Hill Drive, Port Washington, New York 11050. This company manufactures Technamation 35mm motion slides used in conjunction with the Kodak Motion Adapter (see reference). These slides, involving the use of a revolving polarizing disk, show amazing effects of motion to simulate flowing current, lightning, turning wheels, vibrating waves, and the like. They may be employed for display purposes, industrial and commercial application and can be custom made to meet special needs. One educational adaptation is for a Science in Motion program for primary and intermediate grades issued by American Book Co., 450 West 33rd Street, New York, N. Y. 10001.

Bibliography

Aristotle. *De Coloribus.* London: Oxford, Clarendon Press, 1913.

Belmont, I. J. *The Modern Dilemma in Art.* New York: Bernard Ackerman, 1944.

Birren, Faber. "The Emotional Significance of Color Preference." *American Journal of Occupational Therapy,* March-April, 1952.

———. "The Effects of Color on the Human Organism." *American Journal of Occupational Therapy,* May-June, 1959.

———. *Color Psychology and Color Therapy.* New Hyde Park, N. Y.: University Books, 1961.

———. *Color—a Survey in Words and Pictures.* New Hyde Park, N. Y.: University Books, 1963.

———. "Color It Color." *Progressive Architecture,* September, 1967.

———. *Light, Color and Environment.* New York: Van Nostrand Reinhold Co., 1969.

———. "Psychological Implications of Color and Illumination," *Illuminating Engineering,* May 1969.

Castel, R. P. *L'Optique des Couleurs.* Paris: Chez Briasson, 1740.

Cheney, Sheldon. *A Primer of Modern Art.* New York: Liveright, Inc., 1966.

Chevreul, M. *The Principles of Harmony and Contrast of Colors.* New York: Van Nostrand Reinhold Co., 1967.

Cohen, Jozef, and Gordon, Donald A. "The Prevost-Fechner-Benham Subjective Colors." Psychological Bulletin, March, 1949.

Cohen, Sidney. *The Beyond Within.* New York: Atheneum Press, 1964.

Ellinger, E. F. *Medical Radiation Biology.* Springfield, Ill.: Charles C. Thomas, 1957.

Evans, Ralph. *An Introduction to Color.* New York: John Wiley & Sons, 1948.

Field, George. *Chromatics.* London: David Bogue, 1845.

Goethe, Johann Wolfgang von. *Theory of Colours* (Farbenlehre), translated by Charles Lock Eastlake. London: John Murray, 1840.

Goethe's Color Theory. Arranged and edited by Rupprecht Matthaei. New York: Van Nostrand Reinhold Co., 1971.

Gregory, R. L. *Eye and Brain.* New York: World University Library, 1967.

———. *The Intelligent Eye.* New York: McGraw-Hill Book Co., 1970.

Heron, Woodburn, Doane, B. K., and Scott, T. H. "Visual Disturbances After Prolonged Perceptual Isolation." *Canadian Journal of Psychology,* March 1956.

Huxley, Aldous. *The Doors of Preception.* New York: Harper & Row, 1963.

"Intermedia: Tune In, Turn On—and Walk Out." *New York Times Magazine,* May 12, 1968.

Jung, Carl. *The Integration of the Personality*. New York: Farrar & Reinhart, Inc., 1939.

Kandinsky, Wassily, *The Art of Spiritual Harmony*, translated by M. T. H. Sadler. Boston: Houghton, Mifflin Co., 1914.

Karowski, Theodore F., and Adbert, Henry S. "Color Music." Psychological Monographs, Vol. 50, No. 2, 1938. Columbus, Ohio: Ohio State University.

Katz, David. *The World of Colour*. London: Kegan Paul, Trench, Trubner & Co., 1935.

Klein, Adrian Bernard. *Colour-Music the Art of Light*. London: Crosby Lockwood & Son, 1930.

Klüver, Heinrich. *Mescal and Mechanisms of Hallucinations*. University of Chicago Press, 1966.

Leary, Timothy, with Ralph Metzner and Richard Alpert. *The Psychedelic Experience*. New Hyde Park, N. Y.: University Books, 1964.

Leiderman, Herbert; Mendelson, Jack H.; Wexler, Donald, and Solomon, Philip. "Sensory Deprivation." *Archives of Internal Medicine*. American Medical Association, February, 1958.

Luckiesh, M. *Color and Its Applications*. New York: Van Nostrand Co., 1921.

———. *Visual Illusions*. New York: Van Nostrand Co., 1922.

"Luminal Color." *Time Magazine*, April 28, 1967.

Masters, Robert E. L., and Houston, Jean. *Psychedelic Art*. New York: Grove Press, 1968.

Michelson, A. A. *Light Waves and Their Uses*. University of Chicago Press, 1903.

Newton, Sir Isaac. *Optiks*. New York: McGraw-Hill Book Co., 1931.

Ostwald, Wilhelm. *Colour Science*. London: Winsor & Newton, 1931.

Ott, John N. "Effects of Wavelengths of Light on Physiological Functions of Plants and Animals." *Illuminating Engineering*, Vol. 60. April, 1965.

———. "Environmental Effects of Laboratory Lighting." A paper presented before the American Association for Laboratory Animal Science, Las Vegas, Nev., October 21, 1968.

Photobiology, March 1957. A publication of the Division of Biology and Medicine, U. S. Atomic Energy Commission.

"Psychedelic Art." *Life Magazine*, September, 1966.

Rimington, A. Wallace. *Colour-Music, the Art of Mobile Colour*. New York: Frederick A. Stokes Co., 1911.

Rubin, Herbert E., and Katz, Elias. "Auroatone Films for the Treatment of Psychotic Depressions in an Army General Hospital." *Journal of Clinical Psychology*, October, 1946.

Sensory Deprivation. A Symposium held at Harvard Medical School. Cambridge, Mass.: Harvard University Press, 1965.

Spalding, J. R., Archuleta, R. F., and Holland, L. M. "Influence of the Visible Spectrum on Activity in Mice." A study performed under auspicies of U. S. Atomic Energy Commission. Submitted for publication in *Laboratory Animal Care Journal*, 1968.

Stein, Donna M. *Thomas Wilfred: Lumia, A Retrospective Exhibition*. Washington, D.C.: The Corcoran Gallery of Art, 1971.

Unique Lighting Handbook. Barrington, N. J.: Edmund Scientific Co.

Van der Veen, R., and Meijer, G. "Light and Plant Growth." Elindhoven, Netherlands: Phillips Technical Library, 1959.

Vernon, M. D. *A Further Study of Visual Perception*. Cambridge, England: Cambridge University Press, 1954.

———. *The Psychology of Perception*. Harmondsworth, Middlesex, England: Penguin Books, 1966.

Williams, Rollo Gillespie. *Lighting for Color and Form*. New York: Pitman Publishing Corp., 1954.

Wurtman, Richard J. "Biological Implications of Artificial Illumination." *Illuminating Enginnering*, Vol. 63, October 1968, p. 523.

Index

Page numbers in Italic type indicate illustrations